Joseph Squire, Eugene Allen Smith

Report on the Cahaba Coal Field

Joseph Squire, Eugene Allen Smith

Report on the Cahaba Coal Field

ISBN/EAN: 9783337064204

Printed in Europe, USA, Canada, Australia, Japan

Cover: Foto ©berggeist007 / pixelio.de

More available books at **www.hansebooks.com**

GEOLOGICAL SURVEY
—OF—
ALABAMA.
EUGENE ALLEN SMITH, Ph. D., State Geologist.

REPORT

ON THE

CAHABA COAL FIELD,

BY

JOSEPH SQUIRE, M. E.,

ASSISTANT IN CHARGE OF CAHABA FIELD.

WITH

AN APPENDIX

ON THE

Geology of the Valley Regions Adjacent to the Cahaba Field,

BY

EUGENE A. SMITH.

With 31 Figures in the Text, 7 Plates, and a Map of the Cahaba Field and Adjacent Regions.

MONTGOMERY, ALA.:
THE BROWN PRINTING CO., STATE PRINTERS AND BINDERS.
1890.

INTRODUCTORY LETTER.

To His Excellency,
  Thomas Seay,
    Governor of Alabama:

Sir :—I have the honor to transmit herewith, a report on the Cahaba Coal Field, by Joseph Squire. In his letter of transmittal, Mr. Squire has given a short account of the manner in which the map was first begun, and has come finally to be published by the Survey. From this sketch it will be seen that the map and report are the result of more than thirty years work on the part of Mr. Squire, and, at the very low estimate of $1,500 a year, for the compensation of the geologist and for the making of the tests of the seams, they represent an expenditure of at least $45,000 ; the cost of the Survey has been only about one seventh of that sum; the difference has been given to the State by Mr. Squire and those for whom he made the explorations. I think we should not lose sight of these facts. To make the map more complete as to the parts not occupied by the Cahaba Field, I have added the colors showing the distribution of the Geological Formations in the adjacent valley regions, and have appended to Mr. Squire's report, by way of explanation of these colors, a short account of the lithological and other characters of these different geological formations, together with such other matter as seemed necessary to account for the present attitude and positions of the strata of these formations in the valleys. In 1875, 1876, and later in 1882, I have published maps and descriptions showing in a general way, the structure of these valleys, but in the present work, on so much larger scale than any of the previous ones, there was the necessity for much greater detail, and this needed amount of detail concerning the distribution of the various formations, their limits towards each other, and the geological structure, has come chiefly from the notes of

Mr. McCalley, who has devoted the greater part of the last three or four years to the examination of this and the other parts of the State occupied by the rocks of the older formations. The work of Mr. A. M. Gibson in Murphree's Valley, has also been of great service as affording the clew to certain types of geological structure, as will be seen in the body of the report.

I have been somewhat at a loss to determine the best way of exhibiting the distribution of the surface beds of the Tuscaloosa formation in the lower part of the area shown on the map. These beds overlie the older formations in patches, whose exact outlines could not possibly be determined except by instrumental survey, the cost of which would have been out of all proportion to the importance of the information thus to be gained. It must therefore be understood that the map does not pretend to show the exact position and shape of all these overlying tracts—and the absence of definite dotted outlines is intended to indicate this—but only to express the general fact that the Cretaceous beds overlie, and in places completely hide from view, the older geological formations. Where it has been possible to ascertain with certainty, or with a high degree of probability, the distribution of these underlying formations in spite of the covering of Cretaceous, as is easily the case with the Coal Measures, we have so marked it; but in the valley, where several geological formations occur in narrow belts, it has often been quite impossible to trace the continuity of these belts, and thus to determine the structure, hence the unsatisfactory condition of the lower part of the map. In time, and with more numerous observations, we shall probably be able to bring order out of this present confusion.

Of other work completed or in progress in this part of the State, the following statement will not be here out of place.

Mr. McCalley has been engaged for several years upon the examination of the coal of the Plateau region of the State, and of the valleys along which the older geological formations of the State are exposed.

The greater part of this matter is already written up and ready for the printers, and all of it will be ready before the

the end of the winter. The Plateau region includes all that part of the Coal Measures in which the coal seams lie high upon the mountains, and well above the general drainage of the country, and occupies parts or all of the following counties: Madison, Jackson, DeKalb, Marshall, Morgan, Blount, Etowah.

The valley region includes the Tennessee valley, the valley of Blount Springs and immediate valley of the Tennessee river above Guntersville, Murphree's Valley, Wills' Valley, Jones' and Roups' Valley, and Cahaba Valley, and the great valley of the Coosa, embracing all the region between Lookout Mountain and the Coosa Coal Field on the west, and the hills of Clay and Cleburne counties on the east. In these are exposed the older geological formations, and in them occur the beds of red and brown iron ore which have played so important a part in the industrial history of the State. In my biennial report to the present General Assembly I have spoken more specifically of the several reports now ready for the printers upon these districts.

Some years ago, the United States Geological Survey undertook, in the interest of the State Survey as well as that of the United States, an investigation, the chief object of which was to make a carefully measured section of a belt about twenty miles wide, extending across the valley region of Alabama. After consultation, we selected a line running northwest and southeast, near the end of Lookout Mountain at Gadsden, as the central line of this section or belt. The investigation was to determine accurately within this narrow belt, the thickness of the strata of the several formations there occurring, together with the variations in the lithological characters of the rocks from place to place, and to determine the geological structure. This particular belt was selected for the reason that all the older geological formations of the State are exposed here, and the geological structure is about as complicated and diversified as it is anywhere else.

The results of this work, which was finished this fall, are embodied in a report by Mr. C. W. Hayes of the U. S. Survey, illustrated by a map and several geological sections. This report will be published as a document of the State

Survey, for which it was specifically prepared, some time during the fall or winter.

It gives me pleasure to acknowledge still further, the obligation of this Survey to Maj. J. W. Powell, the Director of the U. S. Geological Survey, for the very efficient aid which he has also given us in the investigation of the geology of the southern part of the State. Mr. L. C. Johnson, of the National Survey has spent several months during the past year in field work and in the preparation of a report upon some of the newer formations of Alabama. This report was much needed to make complete the account of the geology of the southern part of Alabama, begun by Mr. Johnson and myself jointly in 1883. The publication of this report has been delayed for two reasons—1st, that we might have a suitable map to illustrate it, and 2nd, that this supplementary work might be done.

The report upon the useful and noxious plants of the State—the timber trees, grasses and other forage plants, weeds, &c., promised by Dr. Charles Mohr of Mobile, has not yet been prepared, because of the illness of the Doctor, but I am glad to be able to say that we shall probably get this most useful report some time during the coming year. No one in the country, north or south, is so well fitted for this task as is Dr. Mohr.

Since the publication of the last report of the Survey, the following assistants have been employed upon the work of the Survey: Prof. Henry McCalley, in examination of the iron ore regions of the State; Mr. Joseph Squire, upon the map and report on the Cahaba Coal Field; Mr. A. M. Gibson, upon the examination of Murphree's Valley, and upon parts of the Coosa Valley; Mr. J. L. Beeson, upon the chemical analyses, which are to go with the Cahaba Coal Field report, and with the report on the iron ore region.

It has been found necessary for Prof. McCalley, who has heretofore had charge of the chemical work of the Survey, to devote his time to field work, and the preparation of his reports thereon, and Mr. Beeson was employed to make the analyses during the past year, but arrangements have been made by which Dr. Wm. B. Phillips, Prof. of Chemistry and Metallurgy at the University of Alabama, will hereafter be in charge of this work.

In addition to these assistants who have been employed by the Survey, we have had aid from the U. S. Geological Survey, as already indicated above, in the work of Mr. C. W. Hayes and his assistants, who have spent several seasons in making the measured section spoken of, and in that of Mr. L. C. Johnson, who has devoted to our work about six months of the past year.

Mr. T. H. Aldrich continues as a volunteer, his study of our Cretaceous and Tertiary fossils, and Mr. D. W. Langdon has given about two weeks of his time to us recently.

The topographic work of the U. S. Geological Survey in our State is going on, and will, in the next three or fours years, have been extended over the entire area of the Warrior Coal Field, and we shall then have a good topographic map on the scale of about two miles to the inch, upon which to show the geology of this region. These topographic maps will make admirable base maps for the illustration of the detailed geological work which the Survey now proposes to undertake, and electrotype reproductions of the plates of these maps will be furnished to the State Survey at the cost of making the same.

I have the honor to be,
Yours most respectfully,
EUGENE A. SMITH,
State Geologist.

ILLUSTRATIONS.

			PAGE
Fig. 1	Section of the Mammoth Seam, Henryellen Basin		26
2	Poole Seam, " "		29
3	Little Pittsb'gh Seam, " "		34
4	" " " " "		35
5	Helena or McGill " " "		36
6	Pump Seam, " "		36
7	Eureka Co.'s Slope Seam, Acton "		43
8	Acton Seam, " "		43
9	Wadsworth Seam, Helena "		50
10	" " " "		51
11	Buck Seam, " "		52
12	Black Shale Seam " "		53
13	Little Pittsb'gh Seam, " "		54
14	Helena Seam, " "		57
15	" " " "		57
16	" " " "		58
17	Wadsworth and Whetrock Seams, Cahaba Basin		66
18	Wadsworth Seam, Eureka Basin		71
19	Helena Seam, Dry Creek "		77
20	Gould Seam, Gould Basin		81
21	Helena Seam, Lolley Basin		86
22	" " " " "		86
23	Montevallo Seam " "		87
24	Black Fireclay Seam, Lolley Basin		88
25	Montevallo Seam, Montevallo "		93
26	Helena Seam, Overturned Measures		98
27	Shaft " " "		98
28	Clark Seam, Dailey Creek Basin		107
29	Gholson Seam, " " "		107
30	" " " " "		107
31	Thompson Seam, Blocton Basin		114
Plate 1	Lancashire method. Endless wire rope haulage, Plan and Section to face		120
2	Method of working steep dipping seams, to face		122
3	Diagram of Slope Tram and ground plan of Slope and room roads, to face		124

Plate 4 Section along the Slope and across the Room entrances, to face...126
     5 Section along the Room roads and across the Hoisting Slope to face...126
     6 Section N W and S E from the Warrior to the Coosa Coal Field, to face...162
     7 View of Coal Seam with Cambrian Limestone overlying it, to face...169

Map of Cahaba Coal Field and adjacent regions, in pocket of cover.

PART I.

REPORT ON THE CAHABA COAL FIELD.

—BY—

JOSEPH SQUIRE.

CONTENTS.

		Page
Letter of Transmittal		1
Chapter I.	General Description of the Cahaba Field	3
II.	The Henryellen Basin	20
III.	The Acton Basin	39
IV.	The Helena Basin	47
V.	The Cahaba Basin	61
VI.	The Eureka Basin	68
VII.	The Dry Creek Basin	74
VIII.	The Gould Basin	78
IX.	The Lolley Basin	83
X.	The Montevallo Basin	90
XI.	The Overturned Measures	95
XII.	The Daily Creek Basin	103
XIII.	The Blocton Basin	111
XIV.	On Mining Methods	118

LETTER OF TRANSMITTAL.

HELENA, ALA., July 30th, 1890.

DR. EUGENE A. SMITH,
    State Geologist.

SIR—I have the honor to transmit herewith my report upon the Cahaba Coal Field of Alabama, with map.

A few words respecting the development of the map. into its present form may not be out of place. The first beginnings of the map were made by me in 1859 or 1860, while in the employ of the Alabama Coal Mining Company, when a fairly correct map of the Montevallo basin was made and the outcrop of the Montevallo seam traced by means of transit and chain. A few years later, under the auspices of the Montevallo Coal Mining Company, these surveys were extended beyond the Montevallo, into the Dailey Creek and Lolley basins. After this, in the years 1867-8, the surveys were still further extended and details worked out as a private enterprise at the joint expense of Dr. I. T. Tichenor and myself.

In 1869-70 the central part of the field, including the Helena, Eureka, and part of the Lolley basins, was explored by me for Daniel Pratt and H. F. DeBardeleben. In 1874 for Mr. T. H. Aldrich my explorations were extended from the Montevallo over parts of the Dailey Creek and Lolley basins, and more recently over a good part of the Blocton and Dry Creek basins.

In 1883, I undertook to make for the State survey, a report and map of the Cahaba field. During the period from 1859 to 1883, we had as above described acquired some pretty accurate, though disconnected knowledge of different parts of the Cahaba field, especially of its lower part; since 1883 our work has been to fill in the gaps and work out the details between these different parts of the field, to connect them together and to trace out from one end of the field to the other, the outcrops of the seams and to reveal the complicated structure of the field as a whole, and as the outcome of this work we have the map in its present form. It will, however, not be amiss to say, that during this period from 1883 to the present time, only about three years' work has been done at the expense of the State, the remaining time having been occupied in surveys and explorations in this field for individuals and companies, with the understanding, however, that the results of these surveys should eventually be turned over to the State to be used in the preparation of this map and report. The two, who have in this way contributed most largely to this work, are Truman H. Aldrich and Henry F. DeBardeleben. It would be impossible to overestimate the public spirit and liberality of men who thus freely present to the State for the benefit of all, the information acquired at great expense to themselves.

In the report, I have not gone into much detail in the description of the different parts of the field, for the reason that the map is constructed to show as nearly as possible, every thing that we know concerning the Cahaba field.

Very respectfully,
JOSEPH SQUIRE.

CHAPTER I.

THE CAHABA COAL FIELD.

The Cahaba Coal Field is part of the great belt or Carboniferous measures that commences near the south boundary line of the State of New York, and continuing southwestward, passes through the States of Pennsylvania, West Virginia, Eastern Kentucky, East Tennessee, and through the north half of Alabama.

The Warrior Coal Field is to the northwest of it, and the Coosa Coal Field is to the east or southeast. Springville is near the northwest corner, Montevallo is near its southeast corner, and Scottsville is near its southwest corner.

Along its northwest side and north end, it is bounded by the Sub-Carboniferous measures; these, and the Silurian and Cambrian beyond, separate it from the Warrior Coal Field. On its southeast side it is bounded by the great "fault" that separates it from the Cambrian measures; these and their overlaying Silurian and Sub-Carboniferous measures, separate it from the Coosa Coal Field; all along its south end it is bounded by a "fault" that separates it from a belt of Cambrian and Silurian measures that intervene between the Carboniferous and the "Drift" measures to the South. This fault is the continuation of that just mentioned.

It is a common saying that the whole world is akin; this saying will apply to our Coal Measures in Alabama. The main characteristic rock formations of the Cahaba Coal Measures are the same as those both of the Warrior and the Coosa Field. By first examining the rocks of the lower half of the Millstone Grit at Brock's Gap (this belongs to the Cahaba field), then examine the base of the Millstone Grit immediately South of Reid's Gap Station (this belongs to the Warrior field), then go out on the Columbus & Western (Central) to Thompson's Gap, or to Carr's Gap on the Georgia Pacific, both on Big Oak Mountain and near Leeds, (these last mentioned gaps being in the Coosa Millstone grit,) you will find them all similar. You will find the same rock at the top of Monte Sano, Huntsville, at the top of Lookout Mountain, Chattanooga, and all along the base of the Coal Measures of Walden's Ridge and Sequatchee Valley, Tennessee; you will also find the same rock at the base of all our Alabama Coal Measures wherever they are the country (or surface) rock.

The underlying Sub-Carboniferous limestone is not very thick near Brock's Gap, becomes thicker going Northwards, as is evident at Blount Springs, where Col. Jackson opened his quarry, and to a greater degree still at Huntsville, where it is over 700 feet in thickness. It also shows a great thickness at Chattanooga.

The Coal Measures of the Cahaba Coal field, like those of the Indian Territory, have only one thin ledge of limestone a few feet in thickness, in the whole of the measures; in both places it is arenaceous and near the middle of the measures.* Richard P. Rothwell, Editor of the Engineering and Mining Journal, New York, was the first to discover this ledge some twenty years ago. The almost entire absence of limestone in our Coal Measures is one of the main points of difference between them and those of the Northern and Western States. Another great peculiarity in our Alabama Coal Measures, in which they differ from anything seen by the writer in the United States, England, Scotland, Wales and the Continent of Europe, is that the great conglomerate of our Coal Measures is at the top of the series.

The five hundred feet of measures above the Montevallo seam are mostly conglomerates or pebbly sandstones (for description of which see chapter on Montevallo basin).

I have no knowledge of any similar case except the Coal Measures near Sydney, Australia, where the top rock of their measures is an immense conglomerate, still larger than ours.

The resemblances between our Coal Measures and those

* A ledge of limestone similar to that described by Mr. Squire is found also in the Warrior Coal Field.   E. A. S.

of other regions, are closest along the lines of latitude. The Coal Measures of Arkansas, for instance, and the Indian Territory, resemble our measures much more than do those of the Northern and Northwestern States. The aggregate thickness of the Cahaba Coal Measures is 5,525 feet; the Arkansas and Indian Territory Coal Measures have over 8,000 feet, while Illinois and Indiana have only about 700 feet in thickness of measures. Another peculiarity in the Cahaba coal seams is the small amount of sulphur in them. While the miners of Illinois are dulling up half a dozen picks a day on flakes of sulphur, most of our miners in the Cahaba field do not hit a flake of sulphur oftener than once a month. In some of our Cahaba seams a miner could not collect a single pound of sulphur flakes in a month. The cause of this absence of pyrites or sulphur in our Cahaba seams can not be explained.

The old idea that our coal seams have been formed from a tropical forest, composed mostly of a large growth of trees is entirely erroneous. An occurrence that happened over thirty years ago, eradicated those ideas, and convinced me that trees of large growth were the exception, and not the rule; at that time it became part of my duty to test and examine a thin seam for a distance of forty (40) miles, and I found its maximum thickness six inches, with a minimum of two inches; this fact and the associated fossils connected with it, convinced me that the vegetation more nearly resembled that of the peat bogs of our day, than anything now existing; in fact, convinced me that the order of formation was from a peat bog to imbedded strata of lignite, and from lignite to the hard bituminous seams of coal now taking our attention, the shrinkage or subsidence of the part of the earth on which they existed, allowing these peat bogs to become covered over with sandy or clayey sediment by the action of water, and a cessation of subsidence, or an elevation, causing the next bog or seam to form. The best evidence of the absence of large trees, (except a few scattered ones,) may be obtained by asking any intelligent old miner that has spent about a third of his time for the last twenty or thirty years underground, to state approximately the number of fossil trees with a diameter over six inches he has seen in the slates and rocks surrounding the coal seams he has mined in his life time experience. In ninety-nine cases out of one hundred he will be able to count them on his fingers; and, when we consider that a coal miner (whether in the room or in the gangway), advances at least two feet per day on an average, or makes an advance of at least three miles in thirty years, with a width of, (using a medium between an eight foot gangway and a twenty-four foot room,) say sixteen feet; his experience should convince any one that the surrounding circumstances at the original formation of a coal seam, resembled those of a peat swamp, instead of a tropical growth of large trees, as the old ideas represented. The evidence is not positive that the climate was tropical at all, but rather that it was mild and of nearly uniform temperature. In evidence of this I will state that the fossil remains of the *Calamites* plant can be found in every ledge from the base of the Millstone Grit to the top of the Montevallo conglomerate, according to my own observation. Now, the living plants most nearly resembling the Calamite, are found in mild and even cool climates. I am informed by men that have been to New Zealand, that the flora of that country more nearly resembles our extinct Carboniferous flora than any they have seen; and the fact is beyond dispute that New Zealand has the mildest climate in the known world; in the Southern part they do not have sun and heat enough to grow our Indian corn. Therefore, following this course of reasoning, that like causes will produce similar effects, we shall be compelled to obliterate our old ideas of a tropical climate with a forest growth of large trees.

Any old coal miner has seen millions of small fossil plants, but I have not met one who has seen a large number of fossil trees.

The Cahaba Coal Field is drained solely by the Cahaba River and its tributaries. This river descends from its northeast end to the south end like a main drain, to which all the creeks and branches on both sides contribute their quota towards making it swell out to such proportions that on leaving the coal field it is large enough for navigation, were it "slackwatered" from the Alabama River to the Coal Field.

Cotton boats are taken down it from the edge of the Coal field, or from Centreville every year. Joseph Lightsey scarcely ever fails taking some boats loaded with cotton down every year; he never attempts it, however, except during high water.

In the south half of the Coal field the principal tributaries on the west side are Schultz's Creek, Caffey's Creek, and Shade's Creek; on the east side of the south half of the Coal field there are Little Cahaba River, Savage Creek, Piney Woods Creek, Beaverdam Creek, and Buck Creek, at Helena. In the north half of the field, the first large tributary of the Cahaba River is the large stream named the East Fork of Cahaba River, or Mill Creek, which joins the river at Parker's Mill, at a point due southeast of Birmingham; then, farther northeast, Black Creek, after draining nearly the whole north end of the Coal Field, joins the river at a point three miles northeast of Henry Ellen. The Cahaba River itself, coming from the direction of Trussville, cuts through the Millstone Grit of Rocky Ridge and enters the Coal field near Hickman's Mill. The amount of coal ever boated down this river is very small; none at all since the war between the States. George Gardner made an effort before the war, for a Montgomery company, to mine coal on Ugly Creek, and boat it down this river; his boats mostly got wrecked on the shoals, and the enterprise was abandoned.

Steamboats have been up this river at times to Centreville, the county seat of Bibb county, a town on the river a few miles south of the Coal Field.

The United States Government made some improvements on the Cahaba River some years ago, with the object of making it navigable. There are some rock shoals between Centreville and the edge of the Coal Field, but below Centreville, I am informed, there are no shoals more serious than gravel shoals to the Alabama River. The distance from the Cahaba Coal Field to the Alabama River by the meanders of the stream is about a hundred miles.

The principal mountain-forming rocks in the Cahaba Coal Field are the Millstone Grit formation and the Montevallo conglomerate.

The highest and most prominent mountains and ridges in the Coal Field are the following: first towards the northwest is Shade's Mountain, formed of the lower measures of the Millstone grit, and following along the northwest boundary of the Coal Field from Canoe Creek in St. Clair county, to a point three miles west of Scottsville, in Bibb county. This ridge, like all the others in the field, changes its name with the locality: thus, in Bibb county it is known as Sand Mountain; in the lower end of Shelby county it bears the name of Farrington Mountain; it is called Shade's Mountain through most of Shelby and Jefferson, and Rocky Ridge in St. Clair county.

The next ridge to the southeast of Shade's Mountain, and parallel with it almost the whole length of the Coal field and formed of the middle portion of the Millstone Grit, bears the name of House Mountain in the south end of Shelby county, of Pine Ridge in the north end of Shelby and south end of Jefferson county, and of Flat Ridge in the north end of Jefferson county, while all over St. Clair county it is called Blackjack Ridge.

The next ridge to the southeast of the two just described, parallel with them, and formed of the upper ledges of the Millstone Grit, is known by the name of Red or Chestnut Ridge in Shelby and Jefferson counties, and by the name of Grassy Ridge in St. Clair county.

The mountains formed by the Montevallo conglomerate are confined to the lower or south half of the Field; the most prominent being Pea Ridge, which is a flat, wide ridge extending from Lacey Station on the Brierfield, Blocton and Birmingham Railroad to the fork of Big and Little Cahaba Rivers. This ridge owes its high altitude to the presence of the conglomerate and to the fact that the measures are nearly flat. It is the broadest ridge in the field and divides the waters of the Big and Little Cahaba Rivers.

The same conglomerate forms another ridge, a little lower in altitude, over the synclinal of the Dry Creek Basin. This is much less extended than Pea Ridge, but nearly as high as Pea Ridge in its central part.

In the northern end of the Coal Field, in addition to the three prominent ridges of the Millstone Grit already described, viz: Rocky Ridge, Blackjack Ridge, and Grassy Ridge, and lying to the southeast of the last named and parallel with it, is Owen's Mountain, formed of the sandstones and slates overlying the Nunally seam. This mountain is not continuous through the field like the others, but in the northern part it is quite as high and prominent as the Millstone Grit ridges.

Besides the mountains above mentioned, which are formed of the rocks of the Coal Measures of the Cahaba Field, there are a few others lying outside the limits of this field, which deserve mention here as affording prominent and important land-marks to guide the explorer in his examinations of the Cahaba Field.

There are two very prominent mountains to the southeast of the Cahaba Coal Field; the first one is a high and continuous cherty ridge running within half or three-quarters of a mile of the Coal field, along its southeast side, with 'Possum Valley between it and the Coal field. This ridge, formed of the chert of the Silurian formation, bears the name of New Hope Mountain in Shelby County, Mill Ridge in Jefferson County, and in St. Clair County it is known by the name of Pine Ridge, changing to Anderson Mountain at the north end. Beyond this to the southeast is a higher mountain than any yet mentioned—the highest in sight of the Cahaba Coal Field. This mountain is known in Shelby and Jefferson Counties by the name of Big Oak Mountain; in St. Clair County some of the settlers call it the Coosa Mountain; about three miles above Carr's Gap, where the Georgia Pacific passes through it, this mountain acquires an altitude exceeding anything in the neighborhood of the Cahaba Coal Field. This high part of the mountain bears the name of "Bald Rock." Big Oak Mountain is formed of the Millstone Grit of the Coosa Coal Field.

On the northwest side of the Cahaba Coal Field and on the opposite side of Shades Valley is the Red Mountain that contains the thick stratified vein of red fossilliferous iron ore, from which the Birmingham furnaces are mostly supplied. This mountain is a prominent land mark along the northwest side of this Coal Field nearly its whole length; its distance from the top of Shades Mountain varies from a half a mile opposite Blocton to about five miles opposite Bessemer, about three miles opposite Birmingham to about two miles opposite Gate City, Shades Valley spreading out between them all the way.

Beyond Red Mountain to the northwest, on the opposite side of Jones' Valley, the Millstone Grit of the Warrior Coal Field forms a ridge at the southeast border of that field. The above mentioned mountains and ridges are most of them shown on the accompanying map.

There are but few good wagon roads in the Cahaba Coal Field; some of them are county roads and have a number of hands apportioned to work them once or twice a year; others are settlement roads, and are either worked by those living along them, by mutual agreement at times when they become extremely bad, or, as sometimes happens, they are neglected and not worked at all; there are other roads that are never worked in any way, and when they become impassable by the falling of a tree or a washout in the road, they are simply turned to the right or left and the obstacle is thus passed, by adopting a new road bed; many of this class of roads become just bridle paths.

The following is a brief notice of some of the best of the wagon roads in this coal field. Beginning at the north end of it, we find the Branchville and Springville road going by David Owen's place. This road is not much used. Father to the southwest is the Branchville and Trussville road going by Hickman's Mill. To the southwest of this is the road from Moody's Cross Roads going by Rock Spring Church to Trussville. Still father southwest is the Rowan Road from Leeds to Birmingham; this road keeps within a short distance of the Columbus and Western, and Georgia Pacific railroads a good part of the way, crossing the railroads at several places and going by Gate City. Farther to the southwest is the road from Pledger's Mill to Gate City and Birmingham; this crosses the Cahaba River at the Glass Ford. To the southwest of this is the Columbiana and Birmingham road; this crosses the Montevallo and Ashville road in Cahaba Valley, at Rufus DeShazo's place, passing by DeLoach and Company's Grist Mill, crossing the Cahaba River at the Dodd Ford. Father to the southwest is the Helena and Birmingham wagon road ; this one crosses the Cahaba River at the Bain Ford, and crosses Shade's Mountain two and a half miles above Oxmoor. The next road to the southwest is the Helena and Tuscaloosa wagon road; this crosses the Cahaba River at the Lainey Ford going by Shade's Creek Church and Greenpond to Tuscaloosa. Still further to the southwest, and crossing a wider part of the Coal Field, is the Montevallo and Tuscaloosa wagon road; this road goes by Boothtown, crosses the Cahaba River at Booth's Ferry or Booth's Ford, joining the Helena and Tuscaloosa road near Shades Creek Church, thence on to Greenpond and Tuscaloosa. To the southwest of this is the Aldrich, Blocton and Woodstock wagon road, going by the D. Lenholm place ; this road is not much used, but crosses the Cahaba Coal Field at the widest part of it, the distance in an air line across the Coal Field from Aldrich to Thrasher's Mill beyond Blocton, is over fourteen miles. To the southwest of this is the Woodstock and Centreville road, going by Randolph's Mill and River Bend. All of the above wagon roads cross the Cahaba Coal Field, some of them diagonally, others nearly direct across. The Cahaba Coal Field away from the mines, is sparsely settled, making road working a heavy burden on the inhabitants, one of whom, James Lindsey, has, himself, made and kept in order for many years, more than six miles of road, in order to keep up communication with neighboring towns. The surface of the Cahaba Field is very broken and contains but a small percentage of level land, that being mostly river or creek bottom land.

The Cahaba Coal Field has the following railroads within its boundaries ; in the north end of the field is the Columbus & Western Division of the Central of Georgia railroad ; this road runs from Birmingham to Opelika and Savannah.

Near it, and alongside part of the way, is the Georgia Pacific railroad ; this road runs from Birmingham to Anniston and Atlanta. Both the above roads pass through Henryellen and Leeds.

Passing through the middle portion of this coal field is the South and North Alabama Division of the Louisville and Nashville company's main line, from New Orleans to

Louisville and Cincinnati. Connected with this main line is the Birmingham Mineral Railroad, from Helena to Gurnee. This Company have a right or lease to run on the railroad from Gurnee to Blocton.

Farther to the southwest is the Brierfield, Blocton, and Birmingham railroad ; this road runs from a point about a mile southwest of Montevallo to Gurnee and Blocton, the main line continuing on from Gurnee to Bessemer, thence over the Alabama Great Southern to Birmingham.

The Cahaba Coal Mining Company have a railroad from Woodstock to their various mines at Blocton; the main line is about nine miles in length ; their branches to the different mines and side tracks increase their railroad mileage to about eighteen or twenty miles.

The Briarfield Coal and Iron Company have a branch railroad running from their coal mines at Peter's Mines, to the East Tennessee, Virginia, and Georgia Railroad at Brierfield ; this road has a length of two or three miles.

The Montevallo Coal and Transportation Company have a branch railroad running from Aldrich on the Brierfield, Blocton, and Birmingham Railroad to their slope on the Montevallo seam.

The Eureka Company have a branch railroad of about two and a half miles in length from their slope in the Helena seam, to the Louisville and Nashville Company's main line at Helena.

The DeBardeleben Coal and Iron Company have a branch railroad from their No. 3 slope above Henryellen to the Columbus and Western Railroad. The above railroads are all completed and in running order, with the exception of the DeBardeleben Coal and Iron Company's branch and the Brierfield, Blocton and Birmingham line from Gurnee to Bessemer; this is all let out under contract to Aldrich, Worthington and Company, and they are pushing the work forward with five hundred to ten hundred hands.

The above railroads are but a small fraction of what probably will be constructed in this Coal Field in the future ; it will require at least ten times their amount in mileage, to bring the Cahaba Coal Field up to its full working capacity.

The Cahaba Coal Field is sixty-eight miles in length by

an average width of five and eight-tenth miles, and contains a surface area of three hundred and ninety-four and a half (394½) square miles. In my computation of the amount of good, workable coal in this coal field, I have included all workable seams of two feet and upwards in thickness, and have excluded all seams in the vertical Coal Measures of the boundary fault, and those of the interior fault, for they are not workable at present and probably never will be, in either fault. The extreme limit in depth of the lowest seams embraced in my computation, is 4,700 feet vertical. With the above named limitations, this coal field contains an aggregate of 3626 millions of tons of coal (tons of 2,000 pounds), from which the loss or waste in mining will have to be deducted. For the amount of coal in each basin, see the chapters on each particular basin.

There are eleven basins in this coal field, besides the "Overturned Measures" at the south end of the Field.

The horizontal sections on the accompanying map illustrate the structure of nearly all of these basins. At the north end of this Coal Field along the line shown on map from "A" to "B," is the *Adkins Horizontal Section*, giving the structure of the north end of the basin and relative positions of seams. The *Henryellen Horizontal Section* gives the structure of the basin and relative positions of seams along the line from "C" to "D." The *Deshazo Horizontal Section* gives the structure of the basin and relative position of the seams along the line on the accompanying map from "E" to "F." Below this is the *Acton Basin Horizontal Section* along the line from "G" to "H," with relative position of same. The *Helena Horizontal Section* along the line from "I" to "J," gives the structure of the Cahaba Basin and the Helena Basin, with relative position of seams in same.

The *Dry Creek Horizontal Section* along the line on accompanying map from "K" to "L," gives the structure of the Gould Basin, and the Dry Creek Basin, with relative position of seams in each one.

The *Blocton and Montevallo Horizontal Section*, along the east and west line on accompanying map from "M" to "N," gives the structure of the Blocton Basin; also that of the Dailey Creek Basin and that of the Montevallo Basin with the relative position of the seams in each basin.

In the vertical sections represented on the accompanying map, the *Henry Ellen Vertical Section* shows the relative position of the seams and rocks in the Henryellen Basin. The *South and North Alabama Railroad Vertical Section* shows the seams and rocks of the Cahaba Basin, the Helena Basin and the measures of the adjacent territory. The *Dailey Creek Vertical Section* shows the seams and rocks that outcrop in that basin between the interior fault and the Stine seam outcrop. The *Blocton Vertical Section* shows a section of the measures that have so far been explored. There are undoubtedly other seams in the part marked unexplored, that the drill or future explorations will bring to light. The *General Vertical Section* shows the combined information gathered from all parts of this Coal field.

The rocks or Coal Measures of the Cahaba Coal Field have an aggregate thickness of 5,525 feet. For the convenience of miners, exploring students, and others, I have classified these measures into four groups:

(1.) The first or lowest group extends from the base of the Millstone Grit to the top of it, or in other words, to the top of the "shield rock" of Chestnut Ridge or Grassy Ridge, between the Gould outcrop and the Nunnally seam outcrop; I have named this one the *"Millstone Grit Group."*

(2.) The next group above the Millstone Grit Group, extends from the top of the Millstone Grit to the top of the hundred feet of blue micaceous sandstone; I have named this group the *"Micaceous Group".* (There are about 200 feet in thickness of measures between the top of the hundred feet of blue micaceous sandstone and the Wadsworth seam.)

(3.) The group above this extends from the top of the hundred feet of blue micaceous sandstone to the Montevallo seam; I have named this one the *"Productive Group."*

(4.) The fourth or topmost group extends from the Montevallo seam to the top of the measures, (about 500 feet); I have named this one the *"Conglomerate Group."*

The rocks forming the dividing line between these groups are good landmarks all over this coal field, wherever they are exposed.

These four groups are all tinted differently on the accompanying map; the different groups are also shown in their respective colors or tints in the vertical and the horizontal sections on said map. The rocks of the "Millstone Grit Group" are colored the darkest tint; the rocks of the "Micaceous Group" are colored a shade lighter than the lowest group; the rocks of the "Productive Group" are colored a shade lighter than the Micaceous Group, and the "Conglomerate Group" is colored the lightest of all. This arrangement, I hope, will enable any one to see at a glance, the class of measures that come to the surface in different parts of the field.

The Cahaba Coal Field, like the Warrior and Coosa Fields, has a great "fault" along its south and southeast boundaries; this is what miners term an "upthrow" fault. For convenience we have named this the "great boundary fault." Unlike the Warrior Field, this has also a similar fault extending down the middle of the field in its southern half; this we have named the "interior fault." At the southern boundary of the field, west of Montevallo, the measures show two faults, the one corresponding to the boundary fault above mentioned, the other, a mile to a mile and a half north of it, following near the old Coffee and Freeman line, for some eight or nine miles. Between these two faults the Coal Measures, including six workable seams of coal, *have been completely overturned*, and left dipping at an angle of sixty degrees towards the southeast. In addition to these, there is a fault separating the Lolley from the Dry Creek Basin, which I have termed the Piney Woods fault, from its position along a creek of that name; and further north, the Beaver Dam fault, between the Dry Creek Basin and the Eureka Basin, named from Beaver Dam Creek which flows nearly along the line of the fault. Besides these faults there are undulations or waves in the measures producing the shallow synclinals with the almost level measures of the Montevallo, the Lolley, the Dry Creek, Dailey Creek, and Blocton Basins with their separating anticlinals.

Outside the flat and undulating measures just mentioned, and the vertical measures near the faults, the strata of the Cahaba Field show an almost uniform southeast dip.

All these displacements of the strata are such as would have resulted from the action of an immense force coming from the southeast, by which the strata were pushed up into folds or wrinkles, lapped over in many cases towards the northwest, and in other cases, fractured along the tops of the folds, and the beds on one side, the southeast, pushed or slipped up over those on the northwest, as is seen in the great faults named. This displacement in the great boundary fault amounts to ten thousand feet; in the case of the interior fault from seven to fifteen hundred feet. Of course this difference in the altitude of the strata on the two sides of the faults does not now exist, and possibly never did, for denuding forces have been active from the beginning planing off the high places and reducing all to a common level, as may be seen for instance, at Helena, where the Cambrian and upper measures of the Cahaba Field, which in their original position are ten thousand feet or more apart, now rest side by side at the same level on the two sides of the great fault.

The small faults or "hitches" in the measures along the northwest edge of the Blocton basin, also result from the action of the same forces, only these faults are much more limited in width and the amount of displacement much less. From the same causes also result the curving of the ends of the Eureka, the Helena, the Acton, and the Henryellen basins, the gentle undulations of the measures in the anticlinal and synclinals of the Montevallo, Blocton, Dry Creek, Dailey Creek, and Lolley Basins, as well as the general southeast dip of the measures of the field taken as a whole.

Along these faults it is the rule to find the upturned measures on the north and northwest side of the fault, standing at a much steeper angle of inclination than do the older measures on the south and southeast sides, which have slidden upon and over them. This is seen all along the great boundary fault, where the upturned edges of the coal measures stand vertical, hence our name of "vertical measures" to designate them. West of Montevallo, as we have seen above, these measures have been pushed over even *past* the vertical, and completely overturned, so that they dip back southeast at an angle of 60 degrees. In all these cases the Cambrian measures on the south and southeast side of the fault have a slope or dip to the southeast rarely greater than fifty or sixty degrees near the fault, and much less than that a short distance away from the fault. Along the interior fault, the same thing may be noticed, as for instance, along the line of the South and North Alabama Railroad near Helena, where the Wadsworth seam at the North "Y" has a dip to the southeast of 42 degrees, while immediately adjacent to this towards the northwest and just across the line of the fault the measures stand vertical, and beyond these vertical measures, which are here about a quarter of a mile wide, we come to the Wadsworth seam again, carried up by this upthrow to the much steeper dip of 50 or 60 degrees to the northwest. And even along the subordinate faults, such as the Piney Woods, we find the measures north of the fault dipping at an angle of 80 degrees north, while those to the south of the same, dip only 35 degrees to the south.

This displacement of two miles vertical along the great boundary fault, and the complete overturn of a strip of country nine miles in length by over a mile in width west of Montevallo, bear witness to the tremendous force that has been brought to bear against the Cahaba Coal Field.

The Cahaba Field is in the counties of St. Clair, Jefferson, Shelby, and Bibb; the northeastern end being in St. Clair county, the southwestern end in Bibb, and the middle portion in Jefferson and Shelby. The county lines according to recent changes, are shown on the accompanying map.

The rate of dip of the measures of the Cahaba Coal Field varies from flat or perfectly level up to sixty degrees from the horizontal. The wide part of the field contains the largest amount of flat measures. In the Lolley and Montevallo Basins you can travel for miles and find it very difficult to decide (judging by the eye) as to which way the measures are dipping. The Blocton Basin holds a large area of flat or level measures; and the same is true of the Dry Creek Basin, and the north end of the Henryellen Basin.

The measures on the southeast side of the interior fault

generally increase their rate of dip as they approach the fault.

The first regular, systematic underground mining in this coal field, was done at a mine opened in the Montevallo seam at a point about one mile northwest of the Montevallo Coal and Transportation Company's present slope, west of Montevallo about three miles; this was about the year 1856; it was commenced by private individuals, and then the Alabama Coal Mining Company was formed, composed of John M. Moore of Talladega, Judge Cooper of Lowndes county, Dr. Miller of Wilcox county, and others. (This was probably the first underground mining done in this State.)

The demand for coal and iron made by the Confederate Government during the war, gave a new impetus to mining coal in this field, and, and new mines were opened near Helena, between Boothtown, and Gurnee, at Dailey Creek, and at the Montevallo Mines, and also to the southwest of Dailey Creek.

Prior to the war, the demand for coal in the whole state was not over ten or eleven thousand tons per annum. For a number of years after the war closed, the demand for coal in Alabama was not much greater than the above. The demonstration had not then been made, that our coal was suitable for smelting iron.

For a number of years after the close of the war, capitalists refused to risk their money in the then doubtful enterprise of building coke furnaces to decide the case as to whether our coals would smelt our ores or not.

The tendency then was to invest in efforts to make cotton with recently liberated slaves, which generally ended in disaster and loss. Matters remained in this condition after the war between the states ended, until the year 1870, when Henry F. DeBardeleben, with a boldness and enterprise that he has shown in many similar cases since then, launched a hundred thousand dollars into the rebuilding of the partly destroyed Red Mountain iron furnaces at Oxmoor, where it was eventually demonstrated that our coal would smelt our iron ores, a fact that we had long craved to see proved beyond dispute. He displayed still greater enterprise in expending between two and three hundred thousand dollars in the opening up of Pratt Mines and bringing cheap coal and coke into Birmingham. Prior to that, capitalists from all parts of the world had seen something of our mineral wealth, but hesitated to venture upon the experiment to ascertain whether the coals and iron ores of Alabama could be worked together in the furnace well enough to make it profitable. It was well known then to a few, that we had an abundance of good coal and iron ores, but that very essential demonstration to induce capital to come here to invest, we did not have.

In the development of the Cahaba Coal Field, the greatest credit must be given to Truman H. Aldrich and Henry F. DeBardeleben. They have done more than any others to push on the developments and mining enterprises that now dot this coal field; Cornelius Cadle and William F. Aldrich have also contributed largely to the mining development of the field.

In locating some of the coal seam outcrops on the accompanying map, after discovering the seam and being positive of its presence, I found it impossible to ascertain its true relative position to the nearest section lines, and distance to nearest section corners, on account of the settlers in the vicinity being unable to point them out; leaving me no alternative but to approximate its position by the apparent distance to some mountain where the section corners were known to me, or to take the compass and chain and run the section lines off; in some of the most important cases, I adopted the last way, and in others of lesser importance, in fact, in the majority of cases, adopted the first way and approximated their position.

With regard to the continuance, or uniformity in thickness of the coal seams shown on the accompanying map, the future developments by further testing and mining will have to decide. I have given the thickness and location of all the seams of the Cahaba Coal Field as accurately as the knowledge obtained up to this date would enable me to do. The condition of a seam of coal, a single yard beyond its exposure, no one living has positive knowledge of, or can rightfully swear as to its size or its purity.

CHAPTER II.

THE HENRYELLEN BASIN.

The Henryellen basin occupies the north end of the Cahaba Coal Field; it is twenty-eight miles in length by an average width of four and a quarter miles, measuring from the base of the Millstone grit on the north-west side of Rocky Ridge to the great boundary fault in 'Possom Valley, on the south side of the Cahaba Coal Field. Its greatest length is measured from the southwest end of the basin at a point about a mile in a straight line south of the mouth of the east fork of the Cahaba river, where it joins the main stream, to the northeast boundary of the basin as well as of the Coal Field, at the Springville prong of Canoe creek.

This basin contains an area of 119 square miles, and is drained solely by the waters of the Cahaba river and its tributaries; chiefly of the east fork of Cahaba river and the numerous prongs of Black creek. The outcropping of the Millstone grit, having a rate of dip of about nine degrees, and forming what is known in this region as Rocky Ridge, clearly outlines the northwestern and the northeastern boundary of the basin, as well as the boundary of this portion of the Cahaba Coal Field. The great fault in 'Possum Valley separating the Cambrian from the Carboniferous measures, forms its south-eastern boundary. Southeast of said boundary fault, and running parallel with it, is a high, prominent cherty ridge of Silurian age, known near the southwest end of the basin as Mill Ridge, near the middle portion of the basin as Pine Ridge, and near the northeast end of the basin as Anderson Mountain. This prominent ridge can be seen from almost any part of the high ridges in the basin, guiding the eye to the location of the basin (also Coal Field) at its foot. The southwest boundary of the basin passes through sections 28, 34 and 35, township 18, range 2, west, intersecting the Cahaba river at a point a little over a mile in a straight line below the mouth of the east fork of Cahaba river; the wagon road from Caldwell's Mill to Watkin's Gap on Shades Mountain crosses the southwest boundary of the basin in section 26. A line commencing at the Alice furnaces in Birmingham, and run to the southeast, would cut off to the northeast, that portion of the Cahaba Coald Field embraced by the Henryellen basin; said line would intersect the first coal seam at a distance from the Alice furnaces of four and a half miles; continuing said line still further southeast, it would reach the southeast boundary of the Cahaba Coal Field (passing entirely over that portion of the field) at a distance of nine miles from the Alice furnaces.

The boundary of the Henryellen basin may be described as follows: Starting from Birmingham with a due southeast course, the top of Red Mountain is reached at a distance of one and a half miles; Shades creek is crossed at three and a half miles, and the base of the Millstone Grit reached at four and a half miles, at a point two-thirds of the way up Shades Mountain on its northwest side, about three hundred yards from the top of the mountain. The course is thence northeast along Shades Mountain (the base of the Millstone grit following along the northwest side of the mountain), with Shades creek meandering along the Valley to the left at a distance of from half a mile to a mile from the crest of Shades Mountain. After continuing along the mountain for three or four miles the ruins of the old Irondale furnace may be seen about half or three quarters of a mile to the left; and beyond, still following along the Millstone Grit, the cut is soon reached through which passes the Georgia Pacific and Columbus and Western railroads; here the Brock seam may be seen exposed in the side of a ditch on the south side of the railroad. After passing this point, Shades Mountain acquires the name of Rocky Ridge, and is known as such by the settlers in its neighborhood all along to its end at the northeast corner of the Cahaba Coal Field, where it intersects the great southeast boundary fault and disappears.

Leaving the railroad behind and continuing along the Rocky Ridge with the base of the Millstone Grit still close to the left, in the valley, on the left the divide between the head waters of Shades creek and a prong of the Cahaba river is soon passed. Trussville may be seen to the left from a point on the mountain about a mile southwest of the gap where Cahaba river penetrates Rocky Ridge. Hickman's Mill is on the river, a short distance up stream, and Revis's Grist Mill down stream from this gap. Our boundary line continues along Rocky Ridge, following the same direction (the measures having a rate of dip to the southeast of about nine degrees), until arriving at a point one and a quarter miles south of Springville, where the Rocky Ridge with its accompanying Millstone Grit changes its direction and bears to the southeast, the Springville prong of Canoe creek following close along its foot at the northeast side. Opposite to Truss' Mill, on Canoe creek, it will be noticed that the Millstone Grit became vertical and the direction of the boundary of the basin and Coal Field changes and bears from this point due south, merging into the great southeastern boundary fault at a point about one and a half miles north of Odenville. The boundary of the basin and Coal Field then continues along 'Possum Valley, passing close by the DeBardeleben Coal & Iron Company's office and store at Henryellen, keeping along the valley north-west of the high cherty ridge known as Anderson Mountain, Pine Ridge and Mill Ridge, for a distance of twenty-six miles, to what is known as Bragg's school-house or Methodist church.

The southwest boundary of the basin extends from this point northwest across the Cadaba Coal Field, intersecting Cahaba river about a mile in a straight line above Caldwell's mill, thence continuing on to the Millstone grit on the northwest side of the basin at a point due southeast from Birmingham and about four or five miles distant.

There are about six public and settlement roads crossing this basin, along which nearly all the rocks of our Cahaba Coal Measures can be examined, except the great 500 feet conglomerate that overlies the Montevallo seam; though a part of this conglomerate can be seen on the Birmingham and Columbiana road near DeLoach & Company's grist mill. One of the roads leaves the Birmingham and Springville road near the Glenn place, crosses the basin and leads to Branchville in Cahaba Valley. Another road leaves Trussville, taking almost a due east course across the Coal Field, and also leads to Branchville. Another road leaves Trussville, crosses the Field and leads to Moody's crossroads in Cahaba Valley. Another road leaves Gate City and Irondale, and crossing the Georgia Pacific and Columbus and Western railroads at places, leads to the Rowan place in Cahaba Valley; this is called the "Rowan Road." Another road leaves Gate City and Irondale and crosses the basin going by the Glass Ford on Cahaba river, and Pledger's Mill on East Cahaba river, to a point on the Cahaba Valley road about two miles north of Bridgeton. Another road leaves Birmingham, crosses Shades Mountain at Watkin's Gap, crosses Cahaba river at the Dodd Ford, crosses the East fork of Cahaba river just below DeLoach and Company's grist mill and leads to Columbiana, crossing Cahaba Valley southwest of Bridgeton about one mile ; this is called the Columbiana and Birmingham road.

At least nine-tenths of the measures of the entire Cahaba Coal Field are crossed by, and partly exposed, along the above roads. The succession of these measures is as follows : Approaching the basin from the northwest, you pass over the sub-carniferous rocks to the base of Shades Mountain or Rocky Ridge ; ascending said mountain you first arrive at the base of the Millstone grit formation, about three hundred yards from the crest of the mountain. About 150 yards above you pass over a seam of coal known as the Brock seam. This seam is thin and not workable in any part of the Cahaba Coal Field where I have yet seen it, though in the northern part of the Warrior Field, I have seen it of good size, holding even as much as four feet of workable coal of good quality. Above this comes a heavy layer of Millstone grit, which in places is a conglomerate of white sandstone with numerous white pebbles imbedded in it, and in other places, a heavy bedded coarse white sandstone. After descending Shades Mountain or Rocky Ridge, you will find heavy layers of gritty slate, in which is a thin seam of four or five inches of coal. Continuing on in the direction of the dip you will ascend another high prominent ridge known in the south end of the basin as Pine Ridge, or Flat Ridge, and in the north end of the basin, as Black-jack Ridge ; this ridge is formed mostly of heavy beds of the white Millstone Grit Sandstone, overlying the gritty slates and forming a shield, protecting the slates from the action of denuding forces. This sandstone is composed of the same material as the white sandstone in Shades Mountain or Rocky Ridge. After passing over this and arriving at the foot of Flat Ridge or Black-jack Ridge on its southeast side, you will cross the outcrop of a thin seam, generally of about six inches in thickness. You are now at the base of the immense gritty slate formation surrounding the Gould seam. Before arriving at the Gould seam you will notice a pink sandstone which is the bottom rock of the under-seam, ten feet below the Gould. Passing over the Gould seam you will find a yellow and pink sandstone, the roof of said seam, and overlying this an immense layer of gritty slate. Ascending the next prominent ridge, which in this basin is mostly designated as Grassy Ridge, (in other parts of the Coal Field it is known as Chestnut Ridge, Red Ridge, &c.,) you now find the thick beds of gritty slate changing their color and texture to layers of sandstone, then gritty slate, and further up the ridge you find a twenty or thirty foot layer of bluish black slate. On attaining the summit of Grassy Ridge, you find the upper layers of the white Millstone Grit Sandstone; this forms the shield to Grassy Ridge against denuding action on the underlying slates. This upper layer is one of our most prominent landmarks in geological examinations in this part of the coal measures of Alabama. Descending the gentle slope of Grassy Ridge to its foot on the southeastern side, you next pass over a number of beds of sandstone and gritty slates and arrive at the Nunnally seam, with a sandstone roof ; this seam contains about two and a half feet of coal. Thence, in the direction of the dip, passing over various layers of sandstone, slaty sandstone and gritty slates, you arrive at the Harkness Double, or Poole seam.

(For section of Poole seam, see below.)

Continuing on in the same direction, you arrive at a large, hundred feet thick layer of blue sandstone, that is very micaceous; this sandstone is a good land mark to guide in locating the measures of this part of the basin. Above this you find the 15-inch Rusty coal seam, and above it, should be found the Wadsworth.*

Keeping your course along the direction of dip, in going over the next 900 feet of measures, you will pass over the outcrop of seven different seams, varying in thickness from six inches to four and a half feet. (See sections on map for details). You will then arrive at a very coarse, massive sandstone. This is the foot wall, or bottom rock, of the Mammoth seam. This coarse sandstone, in various parts of the Cahaba Coal Field, becomes a conglomerate; visibly so at the Henryellen mines and at a point close to the Gurnee mines in the southern portion of the Cahaba Field.

The Mammoth seam, in the north end of this basin, has an aggregate thickness of over eleven feet of coal, and has the following section measured by myself:

---

* I saw what I considered the Wadsworth seam in the northern part of the basin, but was not quite positive of its identity. Anyhow, this is the proper position for it, and I am convinced that future explorations will expose it, yet it is impossible to say what its thickness will be. Near Helena the Wadsworth is over three feet in thickness and makes a very good coke for the smelting iron.

[*Section of Mammoth seam as it shows at the test made on a prong of Black Creek, near the Rock Springs Church, in the north end of the Henryellen basin in section 26, township 16, S., range 1, E. Direction of strike, N. E. and S. W. Direction of dip, S. E. Rate of dip, 15°*].

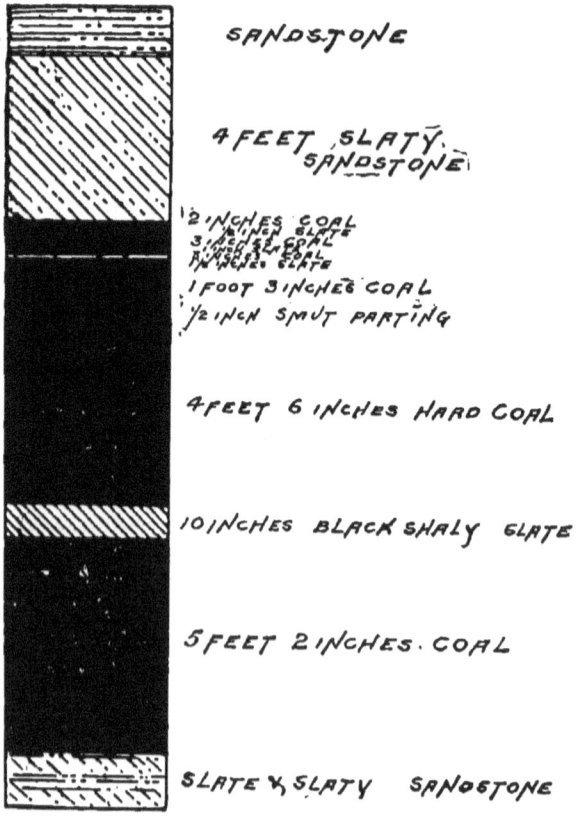

Southwest of this test, a split takes place in the Mammoth, dividing it into two large seams or benches. At Henryellen there are three slopes sunk on the lower or five foot seam, designated as No. 1 slope, No. 2 and new No. 3; the upper or six foot seam they have not begun to work yet. This split is very remarkable on account of the very white sandstone intervening between the two seams in the neighborhood of Henryellen; the thickness of the intervening measures, including the white sandstone in this vicinity, varies from three to thirty-five feet. To the southwest of Henryellen the intervening measures between the two benches increases in thickness to over one hundred feet in places, the sandstone being remarkably white, and very noticeable wherever it is seen. This split in the Mammoth continues on down southwest to the south end of the Cahaba Coal Field, the intervening measures between the two benches varying in thickness from thirty-five feet in the Henryellen neighborhood, to one hundred and eleven feet at Helena, and to ten feet southwest of Gurnee. The Blackshale seam and Buck seam are the names given to the two benches of the Mammoth, near Helena; the Gholson and Clark are the names applied to the same at Gurnee.

Continuing on from the Mammoth along the direction of dip, and passing over three hundred feet in thickness of measures, you cross the outcrop of three thin seams, represented at Helena by the Moyle seam, the Little Pittsburgh, and the Smith-shop seams, you then arrive at the Conglomerate or Thompson seam, which is here five and a half feet in thickness, but impure and slaty and not workable. Continuing along in the direction of the dip, passing over about one hundred and fifty feet of measures, you cross the outcrop of one thin seam, and then arrive at the Helena seam, divided into two layers here, as it is both in the Helena basin, the Dry Creek basin, and the Lolley basin. In the Henryellen basin the Helena seam is double, with four feet of sandstone intervening; the lower layer contains three feet of coal, and the upper layer contains three feet, nine inches of coal; the upper layer is the one on which the Henryellen Company sunk their old No. 3 slope, the coal being of excellent quality.

About sixty-five or seventy feet above the Helena seam is another seam, varying in thickness from two feet to six feet; above this seam there are about a hundred feet of sandstone and slaty sandstone, between it and the vertical fault measures. I have made various measured sections across these vertical fault measures, and could recognize particular seams and rocks, but found them in such a crushed, displaced condition that I could never make the sections match the sections taken in the regular measures of this basin; in the same efforts at the south edge of the Cahaba Coal Field, I met with similar results. When we consider the great disturbance that would inevitably follow the upthrow of 10,000 feet, and the pushing of these measures up to the vertical, we can not reasonably expect all the seams to retain their relative position, so that they can all be identified with same seams in the regular measures of the basin, and even if we could, the fact would have no economic value to the miner, working the seams adjacent, since these vertical seams cannot be profitably worked.

At the south end of this basin, opposite DeLoach & Co.'s grist mill, a steep dip of forty degrees to southeast in the regular measures takes place, bringing additional Coal Measures to the surface and exposing the Montevallo seam and the lower plates of the Montevallo Conglomerate, which can be seen on the road from Birmingham to Columbiana.

At Henryellen, the ledge of conglomerate over the Conglomerate seam shows itself at a point north of the company's store and office, behind the miners' dwellings. It does not show prominently, as it is thin; the pebbles are not abundant, nor large, but they are there. In the Coosa Field southeast of that point, the same plate of conglomerate is reduced to a thickness of two feet. In the south end of the Cahaba Coal Field this plate of conglomerate makes but little better showing than it does in the Henryellen basin. Thin plates of conglomerate are scarcely ever uniform in thickness.

The seams on the east side of the basin, outcropping within two hundred yards of the vertical faulted coal measures, are mostly irregular in thickness, evidently the result of the immense upthrow of the boundary fault. The evenness and regularity in the strike and dip of the coal measures of this basin are extraordinary; I have not noticed any faulty derangement in the interior of the basin except a slight fault showing on section 7, township 18, range 1 west, on Suck Branch and Rocky Branch, though the indications were not serious enough for me to try and work it out. West of Suck Branch, in section 12, close to Henry B. Hanna's house, is an exposure of a seam of coal called in the neighborhood, the Poole seam, of which the following is a section:

[*Poole seam in S. E. ¼ of N. W. ¼, section 12, township 18, S., range 2, W. Direction of strike, N. E. and S. W. Direction of dip, S. E. Rate of dip, 5°*].

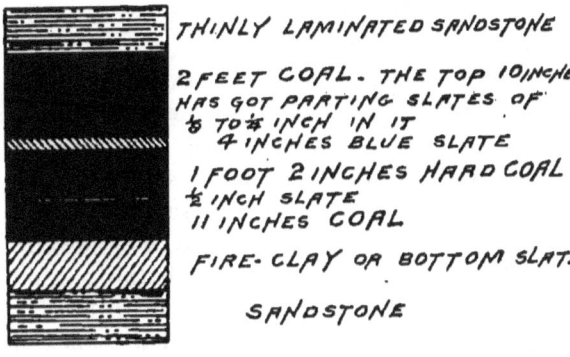

Thinly laminated sandstone

2 feet coal. The top 10 inches has got parting slates of ⅛ to ⅜ inch in it
4 inches blue slate
1 foot 2 inches hard coal
½ inch slate
11 inches coal

Fire-clay or bottom slate

Sandstone

The topography of the Henryellen basin very much resembles that of the portion of the Cahaba Coal Field, near Helena.

The great Millstone Grit formation, here as well as there, forms three high prominent ridges; Rocky Ridge the first one, contains the lowest of the measures, the second or middle one is the Flat Ridge or Black Jack Ridge, the third one is Grassy Ridge. These three are continuous (except where cut through by creeks and branches) all along the northwest side, and the northeast end of this basin. To the southeast of Grassy Ridge, and dividing the waters of Far Black Creek from Middle Black Creek, there is another high ridge that is designated by the settlers in its neighborhood as Owen's Mountain. This ridge follows parallel with Grassy Ridge all along the north end of the basin.

Dividing the waters of Middle Black Creek from the waters of Near Black Creek is another ridge that has the name of Sandstone Ridge, given it by the settlers. These ridges form the principal features in the northwest half of the Henryellen basin.

In the southeast half of the basin the ridges are generally low, the land mostly undulating; the most prominent land mark to be seen from this part of the field being the high cherty ridge, just outside of the basin, and following parallel with the boundary of the coal field on that side.

Black Creek, with its numerous prongs, drains the north

half of the basin and empties into the Cahaba River, near the Henryellen Company's No. 3 slope. The southwest half of the basin is drained by the Cahaba River and its tributaries. In 1883 this basin did not have a single mine opened in it on any of its seams.

The DeBardeleben Coal and Iron Company have three slopes sunk on the lower bench of the Mammoth, and are now mining coal with the most approved machinery and appliances, under the skilful management of Mr. Samuel T. Brittle, with Mr. Hugh Howard as superintendent. Two railroads, (the Georgia Pacific and the Columbus and Western, or Central of Georgia), run through the basin to convey away the coal, and there is a fair prospect of another road very soon. The Macon and Birmingham Company, now building a railroad along 'Possum Valley from Gadsden to Montevallo, would develop by means of lateral roads all the southeast side of the Cahaba Coal Field, and would tap more available coal than any railroad in the State of the same length. The rocks of the vertical coal measures of the boundary fault have the same composition and general appearance that they possess in the interior of the basin.

The measures of the Henryellen basin, like all our Alabama Coal Measures, were evidently at one time approximately level, the ferns and peat mosses of that time in the lakes and bogs of that day, were then forming the carbonaceous matter for our present coal seams. The split in the Mammoth shows that after the first five feet or so of the coal had been formed there was a depression of the seam, 100 feet deep, towards Helena, which became filled up with white sand and other materials from external sources; after it had filled up to a level with the two ends, then the other portion of the Mammoth seam was formed on the top of it.

The present inclined position of the formerly horizontal beds of the Henryellen basin is due to the great fault or upthrow of 10,000 feet on the south-eastern boundary of the basin, and to the upthrow of Jones Valley, which gave its present dip to the Millstone Grit and other measures of the northwest side of the basin. Some men look at this matter as mere conjecture, but they are facts, as to the correctness of which there is no manner of doubt.

The rocks on the southeast side of the basin have a steeper rate of dip than those on the northwest side; this is in accordance with the general law applicable to the whole of the Appalachian region from Alabama to New York, which was formulated by the Messrs. Rogers long ago, as results of their surveys in Virginia and Pennsylvania, and adjacent states.

The method of working the coal seam in this basin hitherto used, has been the method termed by miners, "working the seam on the run." For description of this and other methods see the last Chapter.

The basin contains seams that are of good quality for domestic use; others of good quality for coking and iron manufacturing purposes; and others for a first class steam coal, so that the three principal demands for coal can be supplied by this basin. The following is an analysis of the lower bench of the Mammoth seam, taken from a half bushel sample from the top to bottom of the seam from the Henryellen Company's slope No. 1, at Henryellen. Analysis by J. L. Beeson:

Lower Bench of Mammoth Seam at Henryellen.

Moisture	1.531	
Volatile matter	33.785	
Fixed carbon	59.196	Coke 64.684
Ash	5.488	
	100.000	
Sulphur in coal	1.016	
Sulphur left in coke	.371	
Per cent. of sulphur in coke	.574	

The following is an analysis of the upper bench of the Mammoth seam taken from a half bushel sample channelled out of the seam from top to bottom. This is all from the DeBardeleben Coal and Iron Company's slope No. 1, at Henryellen. Analysis by J. L. Beeson:

Upper Bench of Mammoth Seam at Henryellen.

Moisture	1.526
Volatile matter	33.779
Fixed carbon	53.572 } Coke.......... 64.695
Ash	11.123
	100.000
Sulphur in coal	1.057
Sulphur in coke	.509
Per cent. of sulphur in coke	.787

By referring to the map accompanying this report, the location of the three horizontal sections are shown by dotted lines across this basin; the Adkins section in the north end of basin from "A" to "B"; the Henryellen section in the middle of the basin from "C" to "D"; the DeShazo section in the south end of the basin from "E" to "F". These all show the great disparity between the amount of coal measures in the fault vertical coal measures, and the measures of the interior of the basin, demonstrating that it is utterly impossible for the fault vertical coal measures to be a mere plication such as we find in the basins of the anthracite coal field in Pennsylvania. The accompanying map gives the form of the Henryellen basin as accurately as it could be made without taking the transit and chain and surveying the meanders of the boundary; the seams are also located as accurately as could be done without making a special instrumental measurement and location of each outcrop.

These methods are always used by mining companys in making the special surveys of their property, preparatory to mining development, since the success of their enterprise depends largely on the accuracy of the survey and examination ; in fact, capital can not be safely invested in our coal mining operations without first making these special surveys with all the aids that modern science can give for the purpose of acquiring a full knowledge of the location and condition of the minerals to be mined. In my examination, on the other hand, the area was too extensive to allow me to accomplish all these details, and in view of the fact, that in some days of these examinations, a human face was not visible to me from the rising to the setting of the sun, when the only guide to my location was the forms of the creeks and branches, or my apparent distance from some distant mountain of known location, it will easily be understood that absolute accuracy of location of outcrops was impossible without costly instrumental surveys. Even in thickly settled regions it is often impossible to get a section corner pointed out, since even at best, only a small percentage of the inhabitants have any knowledge of these land marks, and where the ownerships have remained for a long time unchanged, these corners are frequently lost sight of entirely. For these reasons the section corners have not been often referred to.

In the Henryellen vertical section on the map, the seams shown are those that I saw or dug to and found; the three horizontal sections above mentioned on said map, showing the Coal measures at three different points, and stretching across the Henryellen basin, represent the seams of coal that I either actually saw, or identified by the characteristic rocks that are near to and associated with them. Some of them I dug to, without making a thorough test, to convince myself of their identity with the seams in the same relative position in other parts of the Cahaba Coal Field, and to note their peculiarities. I would then abandon the test without obtaining a full section of seam, in order to give more time to forming the general sections, and locating the seams, considering that this result would meet the demands of the people of Alabama better than a number of disconnected details. The extent of the work made it compulsory on me to shun details and economize time as well as expense, so as to obtain the most knowledge of the Cahaba field with the least outlay. In many cases, however, I made very accurate locations of many of the seams shown on the accompanying map, by taking the transit and chain in the one hand, and pick and shovel in the other, and making the one locate what the other brought to light, thus giving me a base line on which to locate the others by reconnoitering and computation of their relative distances. This shunning of the details required an effort on my part, as my work in the past had been largely in making special accurate surveys preparatory to the opening of mines and the investment of capital; so if the reader chooses to find fault about the accuracy of the geological examinations, I shall beg he will excuse me, not on the ground of inability, but on the ground of lack of time and means.

The seams of this basin vary in size, condition and surroundings, but not more so than they do ordinarily in other coal fields. Some are larger here than they are in other parts of the Cahaba Coal Field, while others are smaller. I give below two measured sections of the Little Pittsburgh seam at different localities.

[*Little Pittsburgh seam, on Adkins Spring Branch, in section 26, township 16, range 1 east*]:

LAMINATED SANDSTONE

1 FOOT 4 INCHES COAL

10 INCHES WHITISH SLATE

12 INCHES COAL

BOTTOM SLATE

CAHABA COAL FIELD: HENRYELLEN BASIN.

[*Little Pittsburgh seam, in section 19, township 18, S., range 1, west*]:

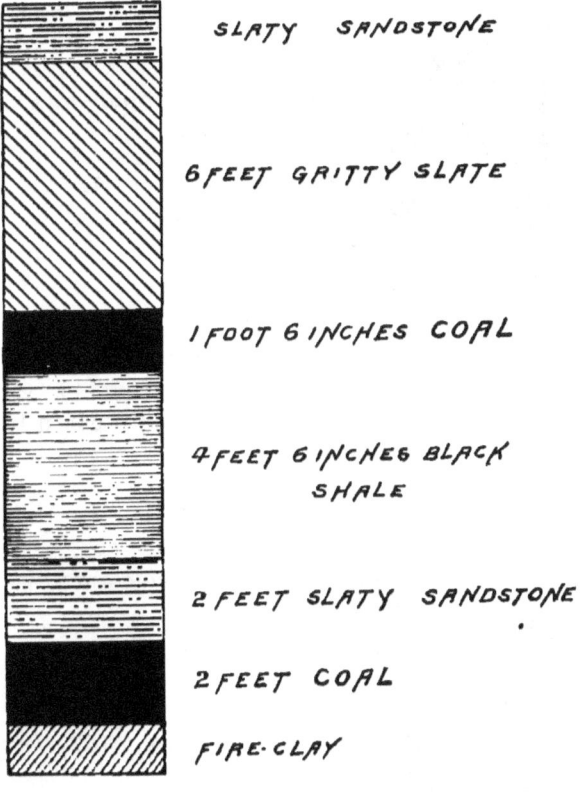

At Henryellen, the old No. 3 slope was sunk on the upper bench of the Helena seam. I did not have the opportunity of seeing it, but Mr. Howard, the superintendent, gave me the following sections:

Sandstone roof.
3 feet 9 inches coal. The slope was in this bench.
4 feet sandstone.
3 feet coal.
Fireclay.

A measured section of the Helena seam in section 26, township 16, range 1 east, is as follows:

[*McGill* or *Helena* seam, in section 26, township 16, S., range 1, east]:

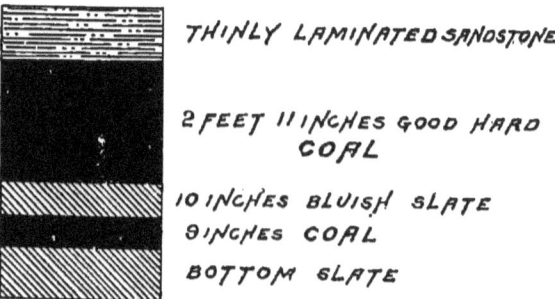

Near Helena is a seam that outcrops under the pump that supplies the coke ovens with water, and named in consequence the Pump Seam, and the following is a measured section of the same seam in section 26, township 16, range 1 east, in the Henryellen basin:

[*Pump seam*, (*under Mammoth*,) in section 26, township 16, S., range 1, east]:

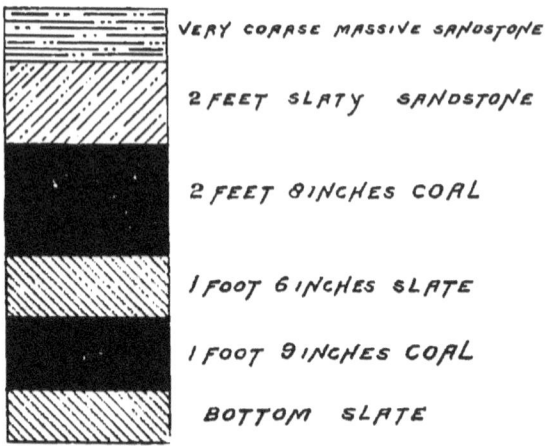

The Henryellen basin contains an aggregate of good workable coal of 881,000,000 of tons of 2,000 pounds. My computation and estimates were made on the basis of including all coal of two feet in thickness and upwards, and all within forty-two hundred feet in vertical depth, but I have made no allowance for loss or waste in pillars or otherwise, in mining.

CAHABA COAL FIELD: HENRYELLEN BASIN.

The surface area of the basin is one hundred and nineteen square miles. The most valuable portion of the basin is on the southeast side; a large amount of the northwestern portion, occupied by Shades' Mountain, or Rocky Ridge, and Black Jack Ridge, contains no seam but the Brock and another thin seam, and as yet I have never seen them of workable size in the Cahaba Coal Field, though the same seams in the northern portion of the Warrior Field and in Tennessee hold four feet and over of good coal.

For a fuller description of the rocks of the Henryellen basin, see vertical section on accompanying map, also Chapter I, giving a general description of all the prominent ledges in the Cahaba Coal Field. For description of the territory surrounding the Henryellen basin see introductory chapter.

The measures of this basin have a varying rate of dip. That portion of it occupied by the Millstone Grit shows a rate of dip generally of from nine to twelve degrees; the measures in sections 6, 7, 8 and 18, in township 16, range 2 east, are nearly flat or level; also in sections 13 and 24, township 16, range 1 east, they are nearly flat; the strata of other parts of the basin have mostly a rate of dip varying from five degrees to twenty-seven without taking into account the fault vertical coal measures of the southeast boundary.

The Coal Measures of the Henryellen basin have a thickness of five thousand feet. In the southern portion of the basin the thickness is a little over that amount, or nearly one mile, counting from the base of the Millstone Grit up to the top of the highest strata of the Coal Measures in the basin.

The following analysis of coke made from the coal of the Mammoth seam at DeBardeleben Coal & Iron Company's Mines, at Henryellen, was made by Alfred F. Brainerd, chemist, Birmingham:

Coke from a Car Load Lot.

Moisture	0.300
Volatile	3.360
Fixed carbon	84.987
Sulphur	0.723
Ash	10.630
	100.000

Analysis of the above Ash from the Mammoth Coke by Alfred F. Brainerd.

Silica	5.000
Alumina	3.500
Oxide Iron	1.921
Lime	0.004
Magnesia	0.003
Sulphur in Ash	0,0002
	10.4282

CHAPTER III.

THE ACTON BASIN.

The Acton Basin at its northeast boundary joins the Henryellen basin, and on its southwest boundary joins the Helena basin and the Cahaba basin.

The principal wagon roads in this basin are the following: the road along 'Possum Valley (part of it is a settlement road), and the Birmingham road that leaves the Cahaba Valley road at Bishop's Mill and the Wilson place, crossing the Cahaba river at the Bain Ford, passing through the Mat Patton place, by Mrs. Bailey's house, where the measures form a synclinal, and crossing Shades Mountain about two and half miles northeast of Oxmoor, thence on to Birmingham. Another road leaves the 'Possum Valley road at William Roy's place, crossing the Cahaba river at the Hubbard Ford, thence on to the top of Shades Mountain, and passing down its northwest side to Oxmoor, thence on to Elyton and Birmingham. Another wagon road leaves the Cahaba Valley road a half a mile above Isaac Johnson's house, going almost due north by Caldwell's mill and Watkin's Gap to Birmingham.

The area of the Acton Basin is forty two square miles. It is drained by the Cahaba River and its tributaries; Patton's creek and its various prongs on the west side of basin, and by Acton's Mill Creek, Coal Branch and other short branches emptying into the Cahaba River on the southeast side of basin.

This basin is not a simple synclinal; but consists of two synclinals with an anticlinal between in its northern end (opposite Oxmoor); the result being a widening of the basin at this point, (see accompanying map). The boundary of this basin may be described as follows: Leaving the L. & N. railroad at Brock's Station (near Brock's Gap) and going due east about three quarters of a mile to the base of the

Millstone Grit, and following this, the course is first north by a few degrees east, along Shades Mountain, keeping Shades Creek and the L. and N. railroad in view on the left all along, passing the large peach orchard owned by Mr. Howell of Cincinnati, and leaving the John McClintock house to the right of the course. In the southeast of section 21, township 19, range 3 west, the course changes nearly due north along the base of the Millstone Grit, the Judge Morrow orchard and vineyard lying distance to the right; this course is followed up to the Hale place. Here Shades Mountain changes direction; and our course is thence northeast passing Oxmoor, with its furnaces on the left, in plain view at the foot of the mountain, a busy little iron manufacturing town. This northeast course continues along the Millstone Grit to the middle of section 20, township 18, range two west; here the rocks are found in irregular position, the ridges more disturbed and broken, and the topography more out of its usual shape by reason of the change of dip between the Henryellen and Acton basins. We go thence southeast along the boundary between this and the Henryellen basin through the middle of section 28; thence through the middle of the north half of section 34; thence through the middle of the south half of section 35, all in township 18, range 2 west, crossing Cahaba River at the west side of section 35; thence to the Methodist church near Mr. Bragg's in 'Possum Valley at the great boundary fault that separates the Cambrian from the Carboniferous. The high cherty ridge on the southwest side of 'Possum Valley here acquires the name of New Hope Mountain. Here the course is changed to the southwest, keeping along the fault at the edge of the Coal Field, and along 'Possum Valley, passing close by the Dave Lowry house, about half a mile from the top of New Hope Mountain; also close by Hens. Bailey's house, with Hale Bailey's a short distance to the left; thence along the edge of the fault to half mile post on the north side of section 2, township 20, range 3 west; thence northwest to Cahaba River, opposite mouth of the Bailey Branch, crossing the Cahaba River at this point; thence up the Bailey Branch in a northwest direction to the half mile post on the west side of section 28, township 19, range 3, west, the point of beginning.

The most prominent ridge in this basin is the Shades Mountain on the northwest side of the basin. Shades Mountain, as already stated, is formed chiefly of the lower portion of the Millstone Grit formation. The northwest side of it can be plainly seen from the L. & N. railroad at almost any point from Brock's Gap in Shades Mountain, to Grace's Gap in Red Mountain, the Millstone Grit forming high perpendicular cliffs near the top of the mountain on its northwest side, displaying the grandeur of nature's handiwork to the thousands travelling along the railroad in the valley. Shades Mountain on its southeast side forms a long gradual slope descending to the slaty valley between it and Pine Ridge, the slope being more gentle and gradual in the north end of the basin than it is in the southern portion.

Pine Ridge is the next prominent ridge in importance and follows nearly parallel with Shades Mountain (on its southeast side), the distance from the top of one to the other varying from half a mile at the south end to a mile at the north end of the basin. The valley between the two is mostly gritty slate, the rocks forming the base of Pine Ridge being also gritty slates and slaty sandstones, the cap or shield of the ridge being a thick ledge of the Millstone Grit formation; in a few places Pine Ridge becomes as high as Shades Mountain.

The next ridge of importance is the Red Ridge; this ridge is southeast of Pine Ridge and follows along nearly parallel with it, the Gould seam with its under and overlying immense thickness of gritty slates, occupying the valley between the two; the cap or shield of Red Ridge is the upper portion of the Millstone Grit formation; these three ridges are continuous along the northwestern side of this basin.

The next ridge in importance is a short distance outside of the southeast boundary of the basin, following along the southeast side of 'Possum Valley; this is the high cherty ridge that is given the name (by the settlers along it) of New Hope Mountain. It intersects the South and North

Alabama Railroad about half-way between Helena and Pelham, the railroad going through a gap cut by Buck Creek in said mountain.

Various smaller ridges are formed in that part of the basin known as the Acton settlement, but they are mostly not continuous like those just described, their general trend is along the strike of the seams and parallel with their outcrops. The Cahaba river also, in one part of this basin, in its general course, keeps along the strike of the seams, following the outcrops and slates until it reaches within a half a mile of the southeast boundary of the Coal Field, a point in section 20, township 19, range 2 west. It then makes a turn away from the southeast boundary again.

The location of the synclinal and anticlinal in the northern part of this basin can be best understood by referring to the accompanying map; on the ground, both can be seen on the road from Bain's Ferry to Birmingham, close to Mrs. Thomas N. Bailey's house. The other synclinal next to the southeast edge of the basin can be seen along the same road at a point about a mile south of Bain's Ferry or Ford. On the accompanying map the Acton horizontal section from "C" to "H" will show the relative position, outcrops, and form of the synclinal and anticlinal of this basin.

The Brock and the Gould seams having a very low rate of dip, become level in the anticlinal between the Mat Patton place and the Mrs. Bailey place, then descending into the main part of the basin, the ledges of conglomerate above the Conglomerate seam show on both sides of the basin along the wagon road between Bain's Ford and the Tom F. Bailey place at the edge of the Coal Field.

There has been no mining for coal in this basin up to this date, except two or three test slopes to prove the seams; but when the basin becomes opened up by railroads its coal seams will undoubtedly be developed.

The Eureka Company's test slope seam, of which the following is a section near the surface, (but I am told it becomes thicker at some depth,) is a seam of good coal and can be worked profitably.

CAHABA COAL FIELD: ACTON BASIN.

[*Eureka Company's slope seam in section 18, township 19, S., range 2, W.: rate of dip 21°*].

Some of the other seams are in good condition for working; the Conglomerate seam is larger and better in this basin than it is at Helena. The Acton seam is large but rather impure; the following is a section of it:

[*Acton seam in section 18, township 19, S., range 2, W*].

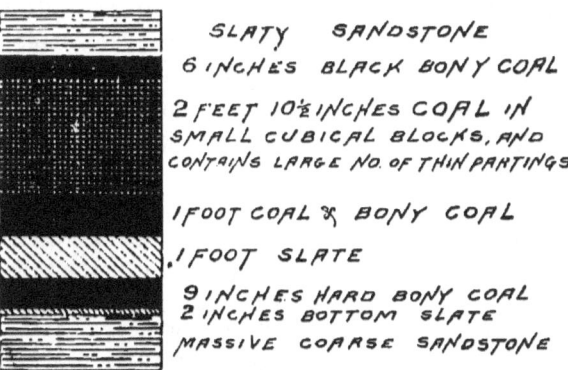

On the wagon road from Caldwell' Grist Mill by Watkin's Gap to Birmingham, at a point about half a mile above Caldwell's Mill, can be seen the flat measures of the anticlinal part of the basin. To the south of said mill about three-qurrters of a mile, the measures have a rate of dip of twenty degrees. The change in the rate of dip can be seen more plainly along the wagon road from Bishop's Mill to Birmingham; the measures becoming more steep as you approach the southeast boundary of the basin, in a similar way to the measures in the south end of the Henryellen basin. The Cambrian measures on the southeast side are the same as those surrounding the southeast side of the Henryellen basin.

The Acton basin is due south from Birmingham; a line from the Union depot, Birmingham, running due south, would cross the top of Red Mountain at a distance of two miles, and intersect the first seam of the basin (the Brock seam) at a distance of five miles; said line continued due south would reach the southeast boundary of the Acton basin at W. Y. Jones' place in 'Possum Valley, at a distance from the Union depot of eleven and a half miles. This due south line would follow very close, almost parallel with the public road that leaves Cahaba Valley at Bishop's Mill, crossing Cahaba river at Bain's Ford and leads to Birmingham.

For relative positions of the seams of this basin, see the *Acton Horizontal Section* from "G" to "H," and the *South and North Railroad Vertical Sections*, both on the accompanying map. The prominent rocks exposed in this basin are very similar to those described in the Henryellen basin. Commencing at the Millstone Grit on the northwest side of Shades Mountain and ascending in the measures by going southeast, you will find an immense ledge of the Millstone Grit, forming all the upper part of the top of the mountain and all the southeast side of Shades Mountain. After passing over this, you will find a thick bed of gritty slate with a thin seam imbedded in it, occupying the valley between Shades Mountain and Pine Ridge. The next ridge (or Pine Ridge) southeast of Shades Mountain has a heavy ledge of the white sandstone of the Millstone Grit formation, for a cap rock or shield; this also underlies the soil on the southeast side. Descending Pine Ridge on the southeast side you will arrive at the immense beds of gritty slate that underlies the Gould seam. After passing over the Gould seam you arrive at the large bed of gritty slate and slaty sandstone that overlies the Gould seam; these gritty slates occupy almost the whole of the Gould Valley, excepting a few thin ledges of pink and red sandstones. On the southeast side of this valley is Red or Chestnut Ridge; this is capped with the upper layer of the white sandstone of the Millstone Grit formation. This cap or shield forms the rocks of the southeast slope of the Red Ridge, descending into the synclinal valley in the north part of the basin and ascending again in the anticlinal farther southeast. Arriving at this point, it will be seen that the micaceous sandstones and slates overlying the Millstone Grit begin to appear, making the upper part of the Millstone Grit a good guide to assist in identifying the upper coal measures. After passing over various layers of sandstone, slaty sandstone, and gritty slate with the Nunnally seam, the "Five Group," and the Harkness seam imbedded in them, we arrive at the large one hundred feet ledge of blue micaceous sandstone. This sandstone is very micaceous and is a great landmark in the identifying of the accompanying coal seams. Overlying this blue micaceous sandstone is the Martin seam, and about one hundred and fifty feet of laminated sandstone interlarded with thin layers of hard block sandstone, some of it breaking out in square blocks. Above this is the Whetrock seam and the overlying Wadsworth seam, and above and including the two just named is the whole productive group of seams up to the Helena seam.

In the northeast corner of section 18, township 19, range 2 west, I found some irregularities of the measures, with indications, however, that the fault was local, or not very extensive. I did not ascertain the extent of it, considering at the time that it did not merit a thorough investigation.

The rate of dip of the rocks of this basin varies from $3°$ or $4°$ in the north part of the basin on the Shades Mountain side, up to $85°$ on the southeast side of the basin adjacent to the great boundary fault.

In the southwest corner of the southwest quarter of section 8, township 19, range 2 west, near the Samuel Acton's house, the rate of dip was found to be $7°$.

On the Mad. Acton land in the southwest quarter of the northeast quarter of section 18, township 19, range 2 west, I found the rate of dip to be $15°$.

In the southeast corner of section 18, township 19, range 2 west, on the T. J. Winfield land, found the rate of dip to be $19°$. On the Acton seam close by, the rate of dip was $21°$; on the Mrs. Jane Acton land in section 20, township

19, range 2 west, the rate of dip was 25°; in the northwest quarter of the northwest quarter of section 36, township 19, range 2 west, on the J. S. Jones' land, the direction of dip is northwest, and the rate of dip 80°. This steep dip is evidently caused by the great upthrow of the boundary fault that is in close proximity at this point. The most level point of this basin is that occupied by the synclinal and anticlinal in the north portion of the basin. The Acton basin is eight and a half miles in length by nearly five miles in average width, and contains an area of forty-two square miles. The amount of good workable coal in it, in seams of two feet thickness and upwards, and less than three thousand six hundred feet in depth, is from careful computation 143,000,000 tons net, making no allowance for waste in mining, loss in pillars, etc.; but this represents the gross amount of coal in the ground.

CHAPTER IV.

THE HELENA BASIN.

The Helena Basin is situated west and north of the town of Helena, and is on both sides of the South and North Alabama Railroad; the greater part being on the northeast side of the said railroad. This basin is bounded on its northwest side by the Interior fault and the Cahaba basin; on its northeast end by the Acton basin; on its southeast side by the great boundary fault and 'Possum Valley; on its southwest end by the Eureka Basin. The length of this basin is three and one-fourth miles, by an average width of three-quarters of a mile. The following is a description of its boundary: Commencing on the South and North Alabama railroad, at a point fifty yards east of the Squire house, at Helena; thence at a bearing of S. 10° W. about half a mile, to a point a little north of the Helena school building; thence, at a bearing of N. 60° W., a distance of one mile, passing to the left of the Holsomback log house in the ridge depression on your way and arriving at a point two hundred yards northeast of the forks of the Tuscaloosa and Birmingham wagon roads, the one fork leading to Lacey's Ford and Birmingham, the other leading to the Lainey Ford and on to Tuscaloosa, both fords being on Cahaba River. You have now arrived at the interior fault, the vertical measures of which are here six hundred yards across. This fault forms the northwest side of the Helena Basin. Thence along the southeast edge of the fault at an average bearing N. 38° E., crossing the South and North railroad at the switch of the north "Y" near Tacoa depot, passing through sections sixteen, nine, ten, three and two, all in township twenty, range three west, and continuing along the southeast edge of the fault, to the point where the interior fault joins the great boundary fault in section two; thence at a bearing of about S. 30° W., along the great boundary fault, on the west side of 'Possum Valley, to the point of commencement on the South and North Alabama railroad at Helena. I have made a more thorough survey and examination of this basin than any other one in this coal field, using the Wye level, the Abney level, the barometer, transit and chain, very liberally; besides making an immense number of test pits with the pick and shovel.

The wagon roads of this basin are the following: There is one at the north end of basin that passes over it for a short distance; this leaves the Ashville and Helena public road at William Roy's house, crosses Cahaba River at the Hubbard Ford, thence on by Oxmoor to Birmingham. Another road leaves the Helena and Ashville road opposite the colored Baptist church at Helena, goes on across the basin to the Cahaba Mines old slope, and to the McClendon and the Driscoll farms. Another road leaves the Helena and Lacey Ford road, and goes on to the Cahaba old slope. A trail or bridle path leaves the Malden Roy house and goes on to the Cahaba old slope at the L. and N. company's bridges over Cahaba River.

The South and North Division of the Louisville system crosses this basin northwest of Helena.

The Gurnee and Blocton Branch of the Birmingham Mineral railroad also runs through a part of this basin and joins South and North near the Scott bridge, or bridge 71.

The Eureka's railroad to their coke ovens and mines, also runs through about three-fourths mile of this basin, joining the South and North railroad near the Scott Bridge at north "Y" of Birmingham Mineral, or Tacoa depot. That portion of this basin situated in sections fifteen and sixteen, is so disturbed by cross faults hitches and distortion of the measures, that it would be very difficult to make a profitable investment in mining in that area, though two-thirds of the basin (that part beyond the cross-fault north of the South and North railroad) are very regular and can be worked profitably. After leaving the South and North railroad going northeast, and advancing along the strike about a quarter of a mile, you will find the measures disturbed by a cross fault. Passing this cross fault, and con-

CAHABA COAL FIELD: HELENA BASIN.

tinuing thence along the outcrop of the seams you will proceed for nearly two miles on measures that have an unbroken regularity, but at the north end the outcrops curve around in a shape much like a fish hook, as shown on the accompanying map; this portion of the basin lies very regular and is well worth the attention of the capitalist and miner. The measures in the southwest end of this basin also curve around in the same fish hook form that they have at the northeast end, as shown by the outcrops of the Helena and conglomerate seams on the accompanying map. The outcrops at both ends of the basin were located by a special instrumental survey by myself.

The causes resulting in the disturbances and irregularity in the measures of this basin are discussed in chapter I, giving the general description of the whole field. Most of the outcrops of the seams of this basin have been carefully surveyed, measurements made, staked off accurately on the surface, and afterwards carefully platted by scale on a map, of which that portion of the accompanying map describing this basin, is the reduced representation.

The great reduction has to some small extent lessened the accuracy. The *South and North Railroad or Helena Vertical Section*, and the *Helena Horizontal Section* "I" to "J," on the accompanying map, will show the relative position of the seams. By referring to the horizontal section, the Helena basin is shown on the right hand side and occupies that portion of the section between the boundary fault vertical coal measures and the interior fault vertical coal measures. The basin on the left hand side is the Cahaba basin, which will be described in the next chapter. The boundary fault on the southeast side of this basin, is an upthrow of ten thousand feet, while the interior fault near the South and North Alabama Railroad has an upthrow of only seven hundred feet, though in the southern part of the coal field this interior fault becomes an upthrow of fifteen hundred feet. The Helena or South and North vertical section gives the seams of both the Cahaba basin and the Helena basin. The coal measures of this basin can be seen most conveniently and to the best advantage, along and near to the South and North Alabama Railroad, between the north

"Y" of the Blocton Mineral Railroad at the Tacoa depot and the Squire house on the main line.

Commencing at said north "Y" and going southeast along the railroad, your first steps will be on the fine grained sandstone underlying the Whetrock seam; you will next find the hard block sandstone thirty or forty feet underneath the Wadsworth; this hard block sandstone is one of the most remarkable rocks for hardness in the whole of our coal measures; it is generally from two inches to six inches in thickness, breaks up into blocks of from two to seven or eight inches across, nearly square. This block sandstone has generally a very pale pea green, or very pale blue color. The first seam you pass over is the Whetrock seam, of about two feet in thickness, dipping to the southeast; all the measures along the South and North Railroad in this basin have a direction of dip to the southeast. Leaving the Whetrock, and passing over forty-seven feet of measures, mostly sandstones, you reach the outcrop of the Wadsworth seam. My oldest pits exposing these two seams, are only a few yards from the South and North Railroad at the point between the north "Y" and the main line. A few years ago the seams could be seen from the railroad, but the wash from the higher ground has covered them up. The following is a section of the Wadsworth seam taken at this point:

[Wadsworth seam in N. E. ¼ of S. W. ¼ in section 16, township 20, N., range 3, W].

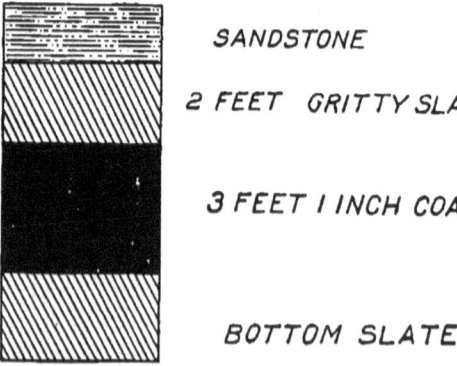

SANDSTONE

2 FEET GRITTY SLATE.

3 FEET 1 INCH COAL

BOTTOM SLATE

Northeast of this in this basin the Wadsworth has a split

in it, as a test made by myself, of which the following section will show:

[*Wadsworth seam in N. E. ¼ of S. E. ¼ in section 9, township 20, S., range 3, W*].

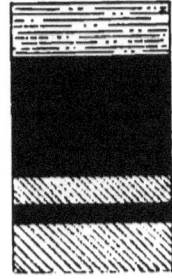

SANDSTONE

2 FEET 10 INCHES COAL

8 INCHES SLATE
5 INCHES COAL
BOTTOM SLATE

Leaving the Wadsworth seam and going southeast, you pass over one hundred and twenty-five feet of measures, mostly coarse red sandstone and hard micaceous grey sandstone; you then arrive at a thin seam of ten inches. Passing over fifty-two feet more of measures, you reach another thin seam of twelve inches; thence passing over one hundred and one feet of fossiliferous grey sandstone and massive grey sandstone, you arrive at the Coke Oven seam, about one and one-half feet thick. This seam is exposed four or five hundred yards south of this point, in the railroad cut west of the old coke ovens, originally built several years ago by the Eureka Company under Mr. Jas. Thomas' directions. The said old ovens are built on the roof of the Coke Oven seam. Passing over forty-four feet of measures you arrive at the Shute seam, outcropping immediately east of said old coke ovens; then passing over three hundred and three feet of measures in the middle of which is a thin seam of about fourteen inches in thickness, you will arrive at the Pump seam. This seam outcrops underneath the steam pump at the wooden bridge or trestle over Buck Creek of the Helena and Blocton Railroad; the outcrop formerly exposed here at this point is now covered up. In this basin its thickness varies from one and a half to seven feet. The last three hundred and three feet of measures are mostly hard micaceous olive colored sandstone, or laminated yellow sandstone. Then continuing southeast, you pass over three hundred and twenty-seven feet of measures, the lower part being mostly slaty sandstone, laminated sandstone, and yellow sandstone, the upper part being a very massive grey or white sandstone which, in other parts of the field, becomes a conglomerate. You then arrive at the Buck seam, which at this point is four feet in thickness, of which the following is a section from actual tests in this basin:

[*Buck seam in N. E. ¼ of N. E. ¼ of section 16, township 20, S., range 3, W*].

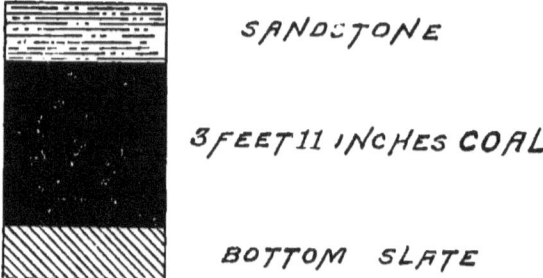

The outcrop of the Buck seam can be seen in the little knoll or point between the south "Y" of the Helena and Blocton Branch of the Birmingham Mineral Railroad and main line near the wooden bridge. The Buck seam is a lower bench of the Mammoth seam, and the same as the seam they are now mining in the No. 1, No. 2, and No. 3 slopes at the DeBardeleben Coal and Iron Company's mines at Henryellen. It is also the same as the Clark seam in the Lolley and Dailey Creek basins. The seam has been worked to a limited extent by the Eureka Company, by means of a tunnel from one of the gangways of their Blackshale slope. The Blackshale slope was south of the South and North Railroad, and in the irregular part of this basin. (*a*).

Continuing southeast along the South and North Railroad, and passing over thirty-five feet of laminated sandstone, you arrive at three streaks of coal, (thin seams of a few inches each,) these thin layers of coal follow the meas-

---

(*a*) I must here state that those conducting and superintending the Eureka Company's work, sank this slope contrary to the advice of the writer, and after their attention was called to the irregularity of that part of the basin.

ures of the Mammoth split, (*b*,) from above Henryellen to Blocton, wherever the writer has seen these rock layers exposed. Then passing over an additional seventy-six feet of fine grained sandstone brings you to the Blackshale seam; this seam is three to three and a half feet thick, on an average, in this basin. This seam is the upper bench of the Mammoth seam, and is also the same seam as the Gholson seam now being worked by the Excelsior Coal Company at the No. 1, No. 2, and No. 3 slopes at Gurnee, on the Brierfield, Blocton and Birmingham Railroad. The Blackshale and the Buck are the Helena equivalent of the Mammoth at Henryellen. The Blackshale is also the same as the Gholson in the Lolley basin and the Dailey Creek basin. From the South and North Railroad to the south end of the field this seam and the Buck occupy an almost continuous valley, along which the engineers have recently located the Helena and Blocton branch of the Birmingham Mineral Railroad, the Buck or Clark being generally near the bottom of, or on the northwest side of the valley, while the Blackshale or Gholson will generally be found on its southeast side, often some distance up the side of the hill. While the Blackshale is six feet at Henryellen and is five feet thick at the old Gholson mine, the average of it, in this basin, as has been proved by actual tests, is not over three and a half feet, yet it is a solid seam of good coal, free from any interlarded layers of slate, smut, or other injurious partings. The following is a measured section of the Blackshale, from a test pit near the South and North Railroad:

[*Blackshale seam, in N. E. ¼ of N. E. ¼ of section 16, township 20, S., range 3, W*].

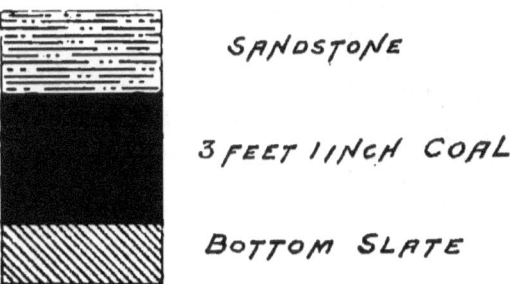

(*b*) The word *split* here refers to the barren strata—sandstones, etc., which come in between and separate the two benches of the Mammoth seam.

The Blackshale seam outcrops a few yards northwest of the south "Y" switch of the Helena and Blocton branch of the Birmingham Mineral Railroad. The old Stevens and Norton slope on the Blackshale, can be seen a few yards northeast of said switch. Leaving the Blackshale seam and continuing southeast along the South and North Railroad, after passing over ninety-seven feet of measures mostly coarse micaceous sandstone, you arrive at a thin seam of about twelve inches, surrounded by rusty slate; the test in this seam is close to railroad on south side. Continuing southeast and passing over one hundred and fifty-six feet of measures, mostly coarse hard grey and red sandstone, you arrive at a double seam, here named the Moyle seam, and varying in thickness from one to three feet, thence southeast, passing over thirty feet of laminated sandstone, brings you to the Little Pittsburgh seam. These two seams outcrop opposite the foundation of an old burnt building on the north side of the South and North Railroad, barely off the right of way; they also outcrop at the south side of Buck Creek where the two old test drifts are seen near the edge of the creek. One of the drifts was made in the Moyle, the other in the Little Pittsburgh ; the wash from the hill has now nearly filled them up. The Little Pittsburgh is also a double seam. At this point, the coal of this seam is of remarkable good quality, but its thickness is too small to justify working. The following is a section of the Little Pittsburgh seam taken from tests made close to this point :

[*Little Pittsburgh seam, in section 16, township 20, S., range 3, W.: rate of dip 35°*].

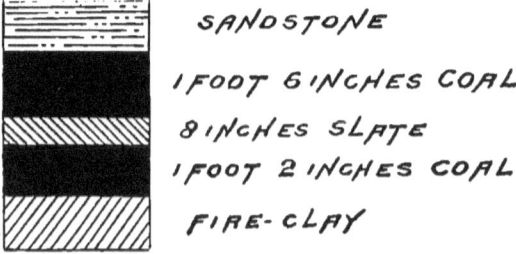

The Little Pittsburgh seam is generally rated as a two and a half foot seam, in this basin. Leaving this seam and continuing along the railroad southeastward, after passing over ninety-two feet of measures mostly hard grey sandstone, you arrive at the Quarry seam. This is a thin seam of one and a half to two feet. Passing over thirty-five feet of additional measures, you will arrive at the Smithshop seam, this is another thin seam of one and one-half feet. The Smithshop seam outcrops in the small ravine or valley immediately southeast of the old quarry; thence from the Smithshop seam southeast, passing over one hundred and seventy-three feet of sandstone, part of it coarse grained, part fine grained, with some massive and some laminated sandstone, you will arrive at the Thompson or Conglomerate seam. The average thickness in the Helena basin, of this seam is from three to five feet, though owing to its close proximity to the great boundary fault, its thickness varies from two and a half or three feet, up to ten or twelve feet. When the seam is in good condition in this basin, it contains from three to five feet of good coal from bottom to top; in places though it becomes interlarded with pockets or layers of what miners call "smut," a black, shiny, soft material that looks very much like coal, and is difficult to keep out of coal on account of its close resemblance, and its not being always at the top of the seam, as the smut that is connected with the Montevallo seam generally is.

The principal defects of the Conglomerate seam in this basin are its roof, (which in places is very treacherous,) its liability to layers of smut, and its irregularity in thickness. Four or five attempts to work this seam in this basin have been made in the past, but in every case have ended in abandoning it, chiefly on the account of the roof and its irregular and defective condition.

The springs in the outcrop of this seam near Buck Creek furnish three varieties of mineral water. On the south side of said creek are two strong chalybeate springs, and from its outcrop on the north side it furnishes a strong alum spring. These waters have been shipped away to some extent, and several invalids have come here to Helena and stayed for the benefit to be derived from these waters. For some classes of bowel diseases they have been highly praised. The Conglomerate seam is the same as the

Thompson and the Underwood, but in the southern portion of the coal field it is much larger, and in better condition than it is in this basin, for description of which see the chapters on the Blocton basin and the Daily Creek basin. In the north end of this basin also, it is thicker and in better condition than it is on the South and North Alabama Railroad.

Leaving the Conglomerate seam and continuing southeast, passing over eighty-four feet of measures, the first twenty feet of which are mostly coarse sandstones, the next fifty feet being a dense conglomerate, some of the pebbles being large enough to make it a puddingstone, and the remaining fourteen feet a hard, coarse sandstone, you come to a thin seam of fifteen inches. This thin seam outcrops in the valley between the Conglomerate ridge and the Helena seam, and the ledge of conglomerate, or the ridge it forms, is an excellent guide and characteristic rock in the identifying and locating of all the other seams in this basin.

The first settlers in this neighborhood gave the ridge the name of Gold Ridge. It may be possible that they presumed that there was gold in it, on account of its containing some quartz pebbles. It is much the highest and most prominent ridge in the basin, and is easily known by the large number of quartz pebbles scattered over it. Leaving the aforesaid fifteen inch seam, and continuing southeastward, you will pass over fifty-two feet of measures, mostly yellow sandstone. This brings you to the Helena seam. That portion of the sandstone immediately under the Helena seam, is fossiliferous, and part of it laminated. The outcrop of the Helena seam is under the railroad trestle between bridge 72 and the Conglomerate ridge. The average thickness in this basin is four to five feet, but in the neighborhood of the South and North Railroad and Buck Creek, a test drift one hundred feet in length close to the creek, demonstrates that its average thickness at this point is not over six inches for the whole length of the drift. The great boundary fault being only about one hundred yards southeast of said drift, sufficiently accounts for the irregularity of the seam at this point. While this seam in the Eureka basin, immediately south of this, has a solid four to four and a half feet of coal without any slates interlarded, in this basin it is usually divided up into two or three layers, as the following measured sections of this seam taken from test pits will show:

[*Helena seam in N. E. 1/4 of N. E. 1/4, section 10, township 20, S., range 3, W*].

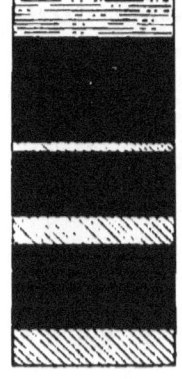

SANDSTONE

2 FEET 6 INCHES COAL

3 INCHES SLATE

1 FOOT 6 INCHES COAL

8 INCHES SLATE

2 FEET COAL

BOTTOM SLATE

[*Helena seam in S. W. 1/4 of S. W. 1/4, in section 2, township 20, S., range 3, W.: rate of dip 30°*].

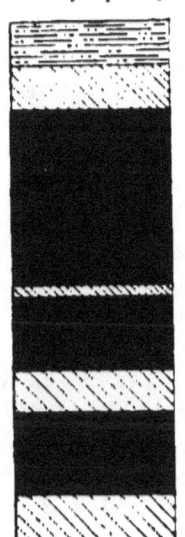

SANDSTONE

GRITTY SLATE

4 FEET 3 INCHES COAL

3 INCHES SLATE

1 FOOT 9 INCHES COAL

1 FOOT SLATE

2 FEET COAL

BOTTOM SLATE OR FIRE-CLAY.

[*Helena seam in N. W. ¼ of S. W. ¼, in section 2, township 20, S., range 3, W.; rate of dip 46°*].

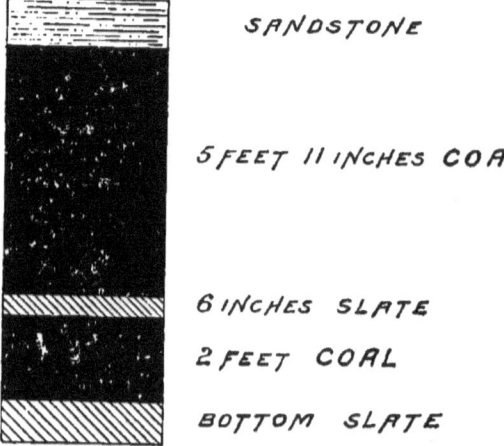

SANDSTONE

5 FEET 11 INCHES COAL

6 INCHES SLATE

2 FEET COAL

BOTTOM SLATE

The coal of the Helena seam ranks very high as a domestic coal, and it is used at present by the Eureka Company for their coke ovens near Helena and their smelting furnaces near Oxmoor, ten miles North of this basin, the large lumps being sold mostly for domestic purposes. The divided condition of the Helena seam is again seen about six miles south of this point in the Dry Creek basin and the Lolley basin; for description of which, see chapters on those basins. From the Helena seam going southeastwards, you pass over ninety-four feet of measures, mostly coarse grey and yellow sandstone and slaty sandstone, forming the high cliff on the south side of the creek opposite the railroad trestle. This brings you to a thin seam of eight inches that outcrops at the pier at southeast end of bridge 72, also in the lane opposite the spring house on the Squire place. This is the uppermost seam outcropping in this basin. Continuing southeastwards, passing over a hundred feet of coarse red and yellow sandstone, containing a large number of *calamites* imbedded in the sandstone in a vertical position as they stood when growing, you will arrive at the great boundary fault separating the Cambrian from the Carboniferous measures, in the grove of willows at the double railroad culvert about three hundred yards north, 73° west, from the Helena depot; the culvert carrying the drainage in the valley south of it to Buck Creek. At this point the fault vertical coal measures are only a few feet across, but north of this at the southeast end of the horizontal section across this basin from "I" to "J," the fault vertical coal measures are more extensive. The direction or strike of the seams and rocks in this basin, along the South and North Alabama Railroad, is about northeast and southwest. The direction of dip about southeast.

The rate of dip varies, and is as follows: In this basin along the South and North Railroad, at the Wadsworth seam, close to railroad, the dip is 42°; at the Pump seam the rate of dip is 40°; at the Blackshale seam, close to railroad, the rate of dip is 38°; at the Smithshop seam on railroad, the rate of dip is 32°; at the Conglomerate seam the rate of dip is 29°, and at the Helena seam the rate of dip is 28°.

The basin is drained by the tributaries of the Cahaba river, Buck Creek making a deep cut through the basin at the south end. The surface area of the Helena basin is two and a half square miles, and its seams, counting all workable coal over two feet in thickness, and to a depth of 2,900 feet, contain 45,000,000 tons (of 2,000 pounds) of coal, making no allowance for waste in mine pillars, or loss in mining. In the foregoing computation I have included the south end of the basin on both sides of the South and North Railroad, though since the recent opening up of the new railroads Helena & Blocton, the Brierfield, Blocton & Birmingham, and the Gurnee & Bessemer, and the Anniston, Syllacauga & Shelby, the said south end has become of more value for manufacturing sites than for mining purposes.

The following analysis of the coal from the Blackshale seam, near Helena, was made by Dr. Otto Wuth, of Pittsburg, from a barrel full of coal from a channelled section of the seam:

```
Water..................................................  .21
Bitumen...............................................33.29
Fixed carbon..........................................64.10
Ash...................................................  2.34
Sulphur...............................................  0.07
```

The following analysis of the coal from the Wadsworth seam, near Helena, was made by Dr. Otto Wuth, of Pittsburg, Pa., from a barrel full of coal from a channelled section of the seam:

```
Water..  ............................................  .42
Bitumen ...........................................31.97
Fixed carbon......................................63.99
Ash..................................................  3.09
Sulphur..............................  ............ 0.53
```

CHAPTER V.

THE CAHABA BASIN.

The Cahaba basin is situated west and northwest of the Helena basin, the interior fault vertical coal measures separate the two.

It is bounded on the southeast side by the interior fault, on the southwest end by the Gould basin, on the northwest side by the sub-carboniferous measures of Shades Valley, and on the northeast end by the Acton basin. The boundary of the basin is as follows: Commencing on the South and North Alabama Railroad, about forty yards south of bridge 70, or Carr bridge, thence southwest along the edge of the fault measures, leaving the Holt house to your right, continuing southwest along the edge of the interior fault, passing close by the northwest corner of section 16, through the middle of section 17 to the middle of the southwest quarter of section 17, thence northwest, crossing Cahaba river and following up Lainey branch to its head, near the northwest corner of section 7, thence over Shades mountain to the base of the Millstone Grit, thence northeast along the base of Millstone Grit through section 6, crossing the South and North Alabama Railroad at Brock's Gap, near the middle of section 32, continuing on northeast to that part of section 28 opposite the head of Bailey's branch, thence southeast down Bailey's branch, crossing the Cahaba river in the south end of section 34, to the vertical coal measures of the interior fault, thence southwest along the northwest edge of the interior fault to the point of beginning on the South and North Alabama Railroad, near the Holt house.

The principal wagon road of this basin is the one formerly called the Montevallo and Elyton road, where, thirty-five years ago, Jemison and Powell used to run their stage coaches, but like the coaches, the road is now very much neglected and out of common use. Said wagon road crosses the Cahaba river at the Lacey Ford, passing under the high railroad trestle in section 5, crossing Shades mountain at Brock's Gap, thence on by Oxmoor to Elyton and Birmingham. On the top of Shades mountain, two other roads branch from this, one going southwest on the top of the mountain towards Gurnee and Blocton, the other one takes a northeast course on the top of Shades mountain and leads to the Morrow orchard, Howell orchard, the Earnest vineyard and the Hale place. Both these last mentioned roads follow along close to the edge of the basin, the roads being but a short distance above the base of the Millstone Grit.

The length of this basin is about three and a half miles from the southwest end to the northeast end, by an average width of two miles, and it contains an area of seven square miles. The amount of good, workable coal in it, in seams over two feet in thickness, amounts to 23,000,000 tons (of 2,000 pounds), at a depth of not over 2,200 feet; in this computation there is no allowance for loss in pillars or waste in mining; about three-fourths of the above 23,000,000 tons are very good coking coals, furnished by the Gould, and Cahaba or Wadsworth seams.

The Cahaba basin is drained by the Cahaba river and its tributaries, Buck creek, Bailey's branch, Black creek, Martin's branch, Lainey branch and others.

The prominent ridges of this basin are Shades mountain on its northwest side, then Pine ridge, near and parallel to the last mentioned, and Red or Chestnut ridge, near and parallel to the other two. The *South and North Alabama Railroad Vertical Section*, and the *Helena Horizontal Section* on the accompanying map, give the relative positions of the seams of this basin; also the form of the basin and its relations to the interior fault and the Helena basin. The horizontal section, showing both basins, is taken along the line shown on map from "I" to "J," said line crossing the South and North Alabama Railroad very near the slope of the South Birmingham Coal and Iron Company, at Sydenton.

The rocks of this basin can be seen to the best advantage along the South and North Alabama Railroad. Commencing at the northwest end of the Brock's Gap cut, the lower part of the Millstone Grit formation can be seen beneath the Brock seam; it has a light bluish tinge. The Brock seam is about one and a half feet thick, the coal being of inferior quality at this point; after passing over forty feet of measures, the Millstone Grit being here of a faint bluish tinge, you come to the seven inch seam; passing over this you will then arrive at the lower part of the two hundred feet of Millstone Grit, you will perceive it here loses its bluish tinge and becomes of a white or grey color, though weathering white; the white pebbled conglomerate is imbedded in this heavy ledge, and though the pebbles in places may not be visible for some distance, they always re-appear again. In general, these pebbles are easily noticed in the Millstone Grit of nearly all our Alabama coal measures. This heavy layer of Millstone Grit forms the shield of Shades mountain, which is the highest in the basin. Crossing over the mountain, in the valley between it and Pine Ridge, you pass over a hundred feet of gritty slate, which you will distinguish from the slate around the Gould, by its containing a greater abundance of rusty parting and bedding planes than the Gould slate does; this slate is of a dirty greenish color. Above this slate is a bluish laminated sandstone. You next arrive at the Millstone Grit of Pine ridge, which can be seen in the railroad cut, locally named the "Teague Cut" in this part of Pine ridge; passing through this you come in sight of the high trestle that stands over the outcrop of the Gould seam and its surrounding slates; you will notice that the gritty slates around the Gould seam are lighter in color than those between Shades mountain and Pine ridge; over the Gould seam is a ledge of yellow and pink sandstone which will help you to locate the seam in almost any part of the Cahaba Coal Field, and over this sandstone is another immense bed of gritty slate. Between said gritty slate and the Millstone Grit of Chestnut ridge, is a ledge of about twenty feet of blue-black slate, quite different from the blue laminated sandstone under the Millstone Grit of Pine ridge. This slate is another guide in identifying and locating the Gould seam. Overlying the blue-black slate is the Millstone Grit of Chestnut ridge; this is the upper layer of

Millstone Grit, and one of its peculiar features is its assuming a more red or pinkish tinge than the layers of Shades mountain and Pine ridge ; it has the same peculiarity in the Warrior Field, which can be noticed along the South and North Alabama Railroad, south of Reid's Gap. Above the Millstone Grit of Chestnut ridge, and both above and below the Nunnally seam, most of the sandstones have a pinkish tinge at their outcrops ; this is a characteristic of this part of the measures. After passing over two hundred feet of measures above the Harkness seam, you will arrive at the lower edges of another great landmark and characteristic rock, the one hundred foot ledge of the blue micaceous sandstone ; a close examination of this ledge will aid you in any investigation of the same series of measures in other parts of the Cahaba Coal Field, (also in Warrior and Coosa Coal Fields.) Passing over this hundred foot ledge, you will find that the sandstones above it are more micaceous than they are below it ; these overlying sandstones acquire a new feature which attaches to most of the ledges immediately below and above the Wadsworth seam—that is, their becoming concretionary, and resembling, when broken the layers or skins of a halved onion ; but the great guide to the identification of the seams in this part of the coal measures, is the large ledge just mentioned of blue micaceous sandstone. The guide to the identification of the Wadsworth seam is the two to six inch ledge of pale blue or green *block sandstone,* which underlies the Wadsworth at a varying distance of from forty to ninety feet. Leaving the Wadsworth seam and continuing southeast, after passing over one hundred feet of measures, you will find a sandstone that is remarkably concretionary in places, but immediately above the Wadsworth is a coarse sandstone that shows very red at the surface. Ascending in the measures to a point one hundred and twenty feet above the Wadsworth seam you will arrive at a hard micaceous grey sandstone containing a thin ten inch seam ; at one hundred and seventy-five feet above the Wadsworth is another thin seam of about twelve inches; about two hundred feet above the Wadsworth is a fossiliferous grey sandstone; about two hundred and fifty feet above the Wadsworth is a massive grey sandstone; above this you will find the Coke Oven seam, and forty-four feet above it the Shute seam, but I do not consider that there is a sufficient area of the two last mentioned seams in this basin to justify preparations for extensive working.

On the northwest side of the Cahaba basin, the rate of dip is very regular, varying from about 15° to 20°; on the southeast side of the basin the rate of dip is much more steep, being mostly from 25° up to 75°.

The Gould seam and the Wadsworth seam are the two principal working seams in this basin, both making a first-class coke; the coke from the Gould seam used to be considered by the foundry men of the State as the best coke that they could get.

The South Birmingham Coal and Iron Company are working the Wadsworth in this basin at Sydenton, by means of a slope driven down southeastwards from the northwest outcrop.

The above mentioned slope, if continued on to the lowest part of this basin, will drain an immense area of the Wadsworth seam. This basin has the great advantage of having the Louisville and Nashville Company's main line (S. and N. A. R. R.) running through the middle of it.

An analysis of the coke recently made from the Wadsworth mine, in the South Birmingham Coal and Iron Company's slope at Sydenton, in this basin, gave the following results:

Analysis of Coke made from the Wadsworth Seam by Alfred Brainerd, of Birmingham, Alabama.

Moisture	0.100
Volatile	2.050
Fixed Carbon	90.183
Sulphur	0.617
Ash	7.050
	100.000

Condition: Good color, ash brick red, specific gravity 1.763.

This is a first rate coke, and one of the best in the Southern States.

The Whetrock seam, or under-seam of the Wadsworth, is thin at this point.

The following is a section of the Wadsworth and Whetrock seams in the Cahaba basin, the Wadsworth being the upper, and separated from the Whetrock by forty feet of measures:

[*Wadsworth and Whetrock seams, at the Carr & Davis slope, in N. W. ¼ of N. W. ¼ of section 9, township 20, S., range 3, W.; direction of strike N., 15° E. from the true meridian, direction of dip S., 75° E., rate of dip 16°*].

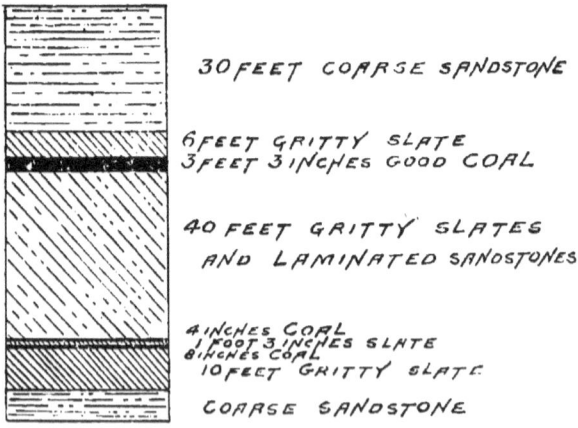

Since the above section was made the South Birmingham Coal and Iron Company, who have bought the property as stated above, have driven the slope further down in the basin and I am informed they found the Wadsworth much thicker than three and a quarter feet.

The Gould seam, I consider after examining it at different points, will average three feet in thickness in this basin; it is an easily mined coal and has a good roof; I have always found it in this basin a solid seam, without any serious layers of slate in it, though I have seen it in the Coosa field with a twelve inch layer of slate in the middle of it. The Gould seam in the Cahaba field bids fair to be worked extensively in the future for the purpose of making a superior quality of coke.

The *South and North Vertical Section* and the *Helena Horizontal Section* (from "I" to "J") on the accompanying map, will show the seams of this basin and their relative position.

The Wadsworth seam in this basin was mined near the railroad bridge during the war by Woodson & Gould, and by various parties since.

Immediately after the war, William Gould opened a drift on the Gould seam at a point about a mile southwest of the high trestle where the Gould outcrop crosses the South and North Alabama Railroad; from this point he supplied the foundaries of Alabama with a superior coke for their cupolas.

For analysis of the Wadsworth coal, see chapter on the Helena basin.

CHAPTER VI.

THE EUREKA BASIN.

The Eureka basin lies southwest of the town of Helena, the north end of it being about half a mile southwest of the Helena depot, on the South and North Alabama Railroad. It is bounded on the north by the Helena basin, on the southeast by the great boundary fault separating the Carboniferous from the Cambrian measures, on the south by the Beaverdam fault, separating it from the Dry Creek basin, and on the northwest side by the interior fault vertical measures.

The following is a description of its boundary: Commencing at the great boundary fault on the east side of the coal field at a point about half a mile southwest of the South and North depot at Helena, thence south by a few degrees west, along the boundary fault leaving Hillsboro fifty yards to your right, leaving R. T. Dunnan's house about a quarter of a mile to your left, continuing along boundary fault until you get nearly opposite Mrs. Peel's house, thence westerly along the Beaver Dam fault, mostly along Beaver Dam Creek, to the half mile post of the south side of section 25, township 20, range 4, west; this brings you to the southeast boundary of the Interior fault measures; thence northeast along the southeast edge of the Interior fault, leaving Lainey Ford six or seven hundred yards to your left, continuing on northeast until you arrive opposite the half-mile post on the east side of section 17, township 20, range 3, west; thence southeast to the point of commencement. Your last course will be nearly parallel with the public road, the road being south or southwest of it.

This basin is drained by the Cahaba River and Beaver Dam Creek and their branches.

The most prominent ridge in this basin is the one that

begins to become high close to Hillsboro, (formed by the roof rock of the Helena seam,) from thence continuing southwest almost over the synclinal of the Eureka basin; this ridge is generally called the Hillsboro Divide, the gorge of Beaver Dam Creek cutting through it. Quite a number of other smaller ridges run parallel with it—the Conglomerate ridge and others.

The length of this basin is three miles, by an average width of one and eight-tenths miles. Its area is five and four-tenths square miles, and it contains, in seams of over two feet in thickness, and less than three thousand feet in vertical depth, 83,000,000 tons of workable coal, (2,000 pounds,) without making any allowance for loss in pillars, or waste in mining.

The form or strike of the measures and coal outcrops in the ends of this basin is quite in contrast to what is seen at the ends of the other basins in this coal field, viz: The measures at the north end are part of them bent sharply around at an acute angle; those at the south end are bent around forming a clearly defined right angle or very near it; the other basins show the measures and outcrops bending around more gradually, some of them forming a half circle or fishhook shape. The lowest seam in this basin workable by slope, is the Wadsworth, the Nunnally seam being too close to the interior fault to allow of it being reliable. The next workable seam above the Wadsworth is the Buck, then immediately above the Buck seam is the Blackshale; both these seams are close to the Helena and Gurnee branch of the Birmingham Mineral Railroad; above these seams and to the southeast of them are the Little Pittsburg seam, the Conglomerate seam, and the Helena seam.

The Eureka Company are now working the Helena seam in this basin by means of a slope driven down from the outcrop to the southeast; said slope is driven down to the synclinal of the basin and is now ascending the opposite dip. The workings in this slope prove the Helena seam to be a good seam of an average thickness of four feet of solid coal, with no slates or impurities except that about two or three inches of the middle of the seam is rather bony; even this burns to an ash along with the other without fail. The coal of this seam ranks high as a domestic coal, but it is now used by the Eureka Company for the purpose of coke-making at their ovens on their branch railroad; said ovens are between the Birmingham Mineral Blocton branch and the Eureka Company's Branch Railroad about a quarter of a mile from Tacoa Station, on the South and North Alabama Railroad. The Eureka Company apply the coke to iron smelting at their Oxmoor furnaces, six miles south of Birmingham.

The Eureka Company's Branch Railroad extends from Tacoa depot, on the South and North Alabama Railroad, to their No. 2 slope, in the Eureka basin, a distance of about two miles. The coke ovens and the houses of the miners are on this branch railroad, between the Louisville and Nashville Company's main line and slope No. 2.

The rate of dip of the measures of the Eureka basin is mostly from $28°$ to $42°$; the exceptions are, the very steep dips on the southeast side of the basin, approaching to the vertical, and the measures of the synclinal which flatten up to a rate of dip as low as $2°$ or $3°$.

The seams of this basin are mostly of good quality; the Wadsworth, a seam of three feet to three and a half feet, yields a very good coking coal, is easily mined, has a good roof, and in the Bee Hive oven makes a first-class coke.

The Buck is a seam averaging about four feet, is a good coal, and will also coke. The Blackshale, a seam of three to three and a half feet, is a very pure, clean seam, makes a good domestic and steam coal, and has a good, hard, safe roof. The Little Pittsburg, a seam of two and a half to three feet, holds an excellent quality of coal for domestic use, but I do not know whether it will make a good coke or not—it is a good steam coal. The Conglomerate seam is also a good coal of from three to five feet in thickness, but liable to layers of smut in the interior of it, so closely resembling coal that none but an expert can well detect it. The Helena is a very good seam of about four feet in thickness, and is also used largely for coking purposes.

The following is a section of the Wadsworth seam in this basin:

[*Wadsworth seam in S. W. ¼ of N. E. ¼, in section 20, township 20, S., range 3, W : rate of dip 38°*].

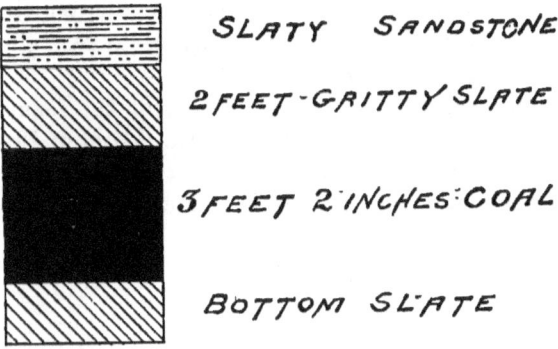

For the relative position of the seams of this basin, see the *South and North Vertical Section* and the *Helena Horizontal Section* from "I" to "J" on the accompanying map.

The only method of working the seams of this basin hitherto practised, has been the method largely used in Pennsylvania of working the coal "on the run," that is, by driving the slope down in the direction of the dip, then driving the gangways horizontally from it, working the rooms up the rise at right angles from the gangways, allowing the coal to run down the room of shutes by its own gravity into the mine cars, a method well suited to all our seams that have a rate of dip of over 40°: (instead of a slope, a drift or vertical shaft can be used.) (*a*).

(*a*) Thirty years ago the writer worked a seam near Montevallo, having a rate of dip of 65° by the same method, and found it suited that rate of dip the very best, but owing to the very steep dip I was compelled to have the miners keep their shutes full up to their "room breasts" to prevent the pulverization of the coal by flinging it violently down an empty or partly empty shute; the coal was loaded in the mine cars at the bottom sufficiently fast, to give the miners working room at the top of the room shute; the run of the coal was checked by curving the bottom of the shute a little, and by using short poles or planks whenever the mine car was full. Very little shovelling was necessary to load the mine cars; part of the room was posted off and lagged for the slate gob; sometimes the coal would scaffold or lodge a considerable distance up the shute, but a shot gun loaded with large buckshot and fired up the shute would loosen it, it being entirely too dangerous for a man to ascend the shute to loosen it.

For all dips of 40° and upwards, the writer considers the above method the best, but whenever the rate of dip becomes low enough to prevent the coal descending the shute of its own accord, then it is not feasible to keep the shute full of coal up to the room breast.

The following four analysis of coal from the seams of the Eureka basin were made by Dr. Otto Wuth, of Pittsburg, Pa, each sample was a barrel full of coal obtained by cutting a channelled section with a pick through the whole seam:

Helena Seam Coal.

Water .... .23
Bitumen 32.53
Fixed carbon 61.26
Ash... 5.85
Sulphur... 0.13

Conglomerate or Thompson Seam Coal.

Water .... .30
Bitumen 31.36
Fixed carban 65.45
Ash... 2.81
Sulphur... .08

Little Pittsburg Seam Coal.

Water .... .18
Bitumen 32.69
Fixed carbon 63.40
Ash... 3.52
Sulphur... 0.21

Moyle Seam Coal.

Water .... .17
Bitumen 31.49
Fixed carbon 60.60
Ash... 7.56
Sulphur... 0.18

The two following analysis were made by J. L. Beeson, from samples obtained from a channelled section of the two seams named:

No. 1.—Helena seam, from the Eureka Company's slope in northern part of S. 29, T. 20, R. 3, W.

No. 2.—Wadsworth seam, from Smith slope of the Eureka Company, S. 20, T. 20, R. 3, W.

CAHABA COAL FIELD : EUREKA BASIN.

	No. 1.	No. 1.
Moisture..	1.669	1.098
Volatile matter	30.541	34.670
Fixed carbon	54.879	59.632
Ash	12.911	4.600
	100.000	100.000
Sulphur in coal	1.141	1.275
Sulphur in coke	.790	.821
Per cent of sulphur in coke	1.666	1.278

CHAPTER VII.

THE DRY CREEK BASIN.

The Dry Creek basin is situated three or four miles southwest of Helena, and northeast of Gurnee. It is bounded on the north by the Eureka basin, on the east by the great boundary fault that divides the Cambrian from the Carboniferous measures, on the south by the Piney Woods fault and anticlinal that separate it from the Lolley basin, on the southwest it is bounded by the interior fault vertical coal measures.

The boundary of the Dry Creek basin is as follows: Commencing at a point about two hundred yards northeast of Lacey depot, on the Brierfield, Blocton and Birmingham Railroad, and going thence along the Piney Woods fault, almost due west, for about two miles; thence along said fault at a bearing of about south $68°$ west, to the southwest corner of section 15; thence northwest to the southeast edge of the interior fault near the northwest corner of section 16, township 21, range 4 west; thence northeastwards along the southeast edge of the interior fault to the half mile post on the south side of section 25, township 20, range 4 west; thence nearly east, or about south $83°$ east, along the Beaver Dam fault to that part of the boundary fault in section 33, township 20, range 3 west, nearly opposite the Mrs. Peel house; thence south by a few degrees west along the boundary fault, passing close by the southwest corner of section 33, leaving the Mrs. Draper house a few yards to the right, passing close by the middle of section 5, then curving around with the boundary fault a little more eastward, to the point of beginning at the boundary fault two hundred yards northeast of Lacey depot.

This basin is drained by the Cahaba river and its tributaries, Beaver Dam Creek, Dave Redding Creek, Peel's Creek, Buzzard Creek, Piney Woods Creek, and Dry Creek.

CAHABA COAL FIELD : DRY CREEK BASIN.   75

The most prominent ridge in this basin is the high ridge over the synclinal of the basin near the southwest corner of section 5, township 21, range 3 west; in this high ridge is seen the roof rock of the Montevallo seam; I saw the outcrop of said seam in the bank of Dry Creek twenty years ago, but it is now covered up by the wash from the hill.

Another prominent ridge in this basin is that known as the "Divide," and it is formed of the roof rocks of the Helena seam, running parallel with the outcrop of said seam from Piney Woods Creek to near the northeast corner of the basin. This ridge, after it leaves the Stinson place, near Piney Woods Creek, runs northeast for about four miles, then turns nearly east to the edge of the coal field opposite the Fountain Wyatt and Mrs. Peel farms.

This basin is five and a half miles in length by an average width of two miles and two-tenths. It contains a surface area of twelve and one-tenth square miles, and contains in workable seams of two feet and upwards in thickness 202,000,000 of tons of coal, (2,000 pounds,) without making any allowance for loss in mine pillars, or waste in mining; this amount of coal is within a limit of 4,300 feet in vertical depth.

The wagon roads of this basin are the two Lindsey roads (made by James Lindsey); one of them runs from his place in the northeast corner of section 3, township 21, range 4 west, bearing southeast through the south half of the basin and joins the Helena and Montevallo wagon road at the Mrs. Lacey place and the Carroll place. The other Lindsey road leaves the Lindsey farm and runs northeast along the strike of the seams to Helena. Another wagon road leaves the Helena and Montevallo road at the Mrs. Peel place and the Fountain Wyatt place, and follows along the top of the Divide ridge down to Piney Woods Creek—this road leads to Gurnee. Another wagon road leaves the William Lacey place in 'Possum Valley and following along the edge of the basin leads to the Ryan place on the "Divide," in the southwest corner of the basin.

The Brierfield, Blocton and Birmingham Railroad extends along the south boundary of the basin for a distance of about five miles.

The Blocton Branch of the Birmingham Mineral Railroad passes through the western portion of the basin for a distance of five miles, extending on to Gurnee, and, having a lease from the Brierfield, Blocton and Birmingham Road from Gurnee to Blocton, the same road is enabled to connect with Blocton.

The principal workable seams of this basin are the Buck seam, Blackshale seam, Conglomerate seam, Helena seam, and the Montevallo seam. The Shute and the Coke seam are in workable condition southwest of this in the Dailey Creek basin, but in this basin, a thorough test along their outcrops will have to decide their condition for mining purposes.

The rate of dip of these measures in this basin varies from $2°$ or $3°$ in that portion south of Dry Creek, to $80°$ at the south edge of the basin next to the Piney Woods fault. The measures on the west or northwest side have an intermediate rate of dip between the dips of the two previously mentioned points.

The *South and North* and the *Dailey Creek Vertical Section* and the *Dry Creek Horizontal Section* from "K" to "L," on the accompanying map, show the relative position of the seams of this basin.

There has been no mining done in this basin except a little outcrop coal dug for blacksmith purposes by the farmers in the neighborhood, no underground work has been done in any part of it up to this date.

There is an immense amount of coal nearly level in this basin with the advantage of two recently constructed railroads, now nearly finished, running through and alongside of it—the Brierfield, Blocton and Birmingham on the south edge of it, and the Birmingham Mineral in the northwest portion of it. This basin has been a wild, sparsely settled country up to about twelve months ago ; two years ago no one lived in the interior of the basin ; at that time the only settlers about it were Mrs. Draper and her son, D. D. Draper, Herve and Burt Carroll on the east boundary of the basin, Columbus Benton on the north boundary, and James Lindsey on the western boundary of the basin. This basin bids fair to become the scene of busy mining operations in the near future.

CAHABA COAL FIELD: DRY CREEK BASIN.

The following is a measured section of the Helena seam at its southern outcrop in section 12, the measures here having a very steep rate of dip:

[*Helena seam in section 12, township 21 S., range 4 W.; direction of strike N. 65° E., S. 65° W. magnetic; direction of dip N. 25° W.; rate of dip 80° from horizontal.*]

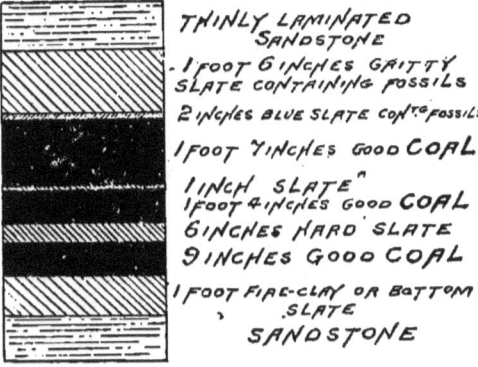

THINLY LAMINATED SANDSTONE
1 FOOT 6 INCHES GRITTY SLATE CONTAINING FOSSILS
2 INCHES BLUE SLATE CONT'G FOSSILS
1 FOOT 7 INCHES GOOD COAL
1 INCH SLATE
1 FOOT 4 INCHES GOOD COAL
6 INCHES HARD SLATE
9 INCHES GOOD COAL
1 FOOT FIRE-CLAY OR BOTTOM SLATE
SANDSTONE

CHAPTER VIII.

THE GOULD BASIN.

The Gould basin is situated to the north of Gurnee, to the southwest of Helena, and on the northwest side of the Cahaba Coal Field; it is bounded on the southeast side by the Interior fault vertical coal measures, on the northeast end by the Cahaba basin, on the northwest side by the Sub-Carboniferous measures of Shades Valley, on its southwest end by the Blocton basin.

The following is a description of the boundary of said basin: Commencing on the northwest edge of the Interior fault vertical measures, near the mouth of Lainey Branch; thence northwest along Lainey Branch to the base of the Millstone Grit at a point a half a mile northeast of Genery's Gap where the Brierfield, Blocton and Birmingham Railroad cuts through Shades Mountain; thence southwest along the base of the Millstone Grit, crossing the Brierfield, Blocton and Birmingham Railroad at the northwest end of the Genery Gap Railroad cut in Shades Mountain; continuing on southwest along the base of the Millstone Grit, Shades Valley being in plain view, leaving the Richard Tyler house and the Squire John Harmon house to your left; from opposite the John Harmon house your course will be more westward, (about 70° west,) continuing along the base of the Millstone Grit, crossing Shades Creek a short distance below the mouth of Roup's Creek, leaving Kimbrall's Mill to your right, until you arrive at a sharp bend in Shades Mountain in the south end of section 3, township 21, range 5 west; from this point southeastwards along the wagon road that leads from Booth's Ferry to Roup's Iron Works, crossing Shades Creek near Shades Creek church, leaving the Miller farm and the flat measures of the Blocton basin to your right; crossing the Cahaba River at Booth's Ferry near the mouth of Lick Creek'; a few yards farther brings you to the Interior fault vertical coal measures; thence northeastwards along the northwest edge of the vertical measures of the Interior fault, crossing Cahaba river again in the southwest corner of section 17, township 21, range 4 west; continuing along the edge of said fault, crossing Ward's Creek, Shaw's Creek, mouth of Hurricane Creek, the two Sandstone branches; crossing Trigger Creek and continuing on to opposite the mouth of Lainey Branch, the point of commencement; this point is about three-quarters of a mile northeast of Lainey Ford.

The Gould basin is drained by the Cahaba river and its tributaries—Shades Creek, Hancock Creek, Ward's Creek, Shaw's Creek, Hurricane Creek, Little Sandstone Branch, Big Sandstone Branch, Trigger Creek and Lainey Branch.

The most prominent ridge in this basin is Shades Mountain; at the southwest end of this basin it is named Sand Mountain. The next one in size and prominence is the one next to Shades Mountain on its southeast side; running parallel with it. This is called Pine Ridge in the northeast end of the basin—but is named House Mountain in the middle of the basin, and Hurricane Ridge in the southwest end of the basin. The next one in size and prominence is Red Ridge. This one, on the South and North Alabama Railroad, is called Red or Chestnut Ridge, and contains the upper measures of the Millstone Grit formation.

These three ridges just mentioned are all parallel with one another from one end of the basin to the other. At the southwest end they become broken. There are other ridges of less prominence, mostly running parallel with those above mentioned. All these ridges are cut by some of the smaller creeks and branches, except Shades or Sand Mountain; this mountain is cut through only in one place, that is at the southwest end of the basin where Shades Creek cuts a gap in it, in its course from Shades Valley to Cahaba river.

The length of this basin is nine and three-quarter miles by an average width of two and two-tenths miles, and it contains a surface area of twenty-one and a half square miles. It contains in seams of two feet and upwards of workable coal, 77,000,000 tons (2,000 pounds), within a limit of 2,500 feet in vertical depth; in this computation no allowance is made for loss in pillars, or waste in mining the coal.

The principal wagon roads in this basin are the Tuscaloosa and Columbiana road—this road enters the southwest end of the basin near Shades Creek church, and continues along the foot of the southeast side of Red Ridge nearly all the way to Lainey Ford where it leaves the basin. The next wagon road in importance is the one at the southwest end of the basin leading from Booth's Ferry to Tannehill Station, on the Alabama Great Southern Railroad. Another wagon road connecting Brock's Station with John Harmon's place and Kimbrall's Mill, leads along the top of Shades Mountain from near Brock's Gap to John Harmon's, there it descends the north side of the mountain and leads to Kimbrall's Mill in Shades Valley. Another wagon road leaves the Columbiana and Tuscaloosa road, where said road intersects Hurricane Creek, follows up the side of Hurricane Creek passing close by Lindsey's old mill and gin joining the road on the top of Shades Mountain at Richard Tyler's. Another wagon road leaves the Tuscaloosa and Columbiana road two or three hundred yards southwest of Lainey Ford, passes through the Horton and Doss places, then through Genery's Gap to Bessemer and Birmingham.

The Brierfield, Blocton and Birmingham Railroad enters the basin at the north end of sec'ion 9, township 21, range 4 west, follows up Ward's Creek, passing through gaps in Red Ridge and House Mountain or Pine Ridge; then passing through the deep cut in Shades Mountain at Genery's Gap; thence across Shades Valley passing through Spark's Gap in Red Mountain and on to Bessemer and Birmingham over the Alabama Great Southern Railroad. This part of the Brierfield, Blocton and Birmingham Railroad extends from Gurnee to its junction with the Alabama Great Southern at a point about three miles southwest of Bessemer. In its course it passes over the outcrop of the Gould seam.

The most important and valuable seam in this basin is the Gould seam; it extends the whole length of the basin. A few years ago, J. L. Davis made a series of tests along the outcrop for about six miles in this basin, and as a result

CAHABA COAL FIELD: GOULD BASIN.   81

of said tests, reported that the average thickness of the Gould seam was about three feet. This seam has the reputation of making a coke equal to the Pocahontas, for iron smelting purposes, and it can be easily mined; probably in the future it will supply a good part of the demand for a superior coke. Twenty years ago it had the best reputation of any in the State, as making a good cupola or iron foundry coke. The Gould seam in this basin is not yet mined, as the Brierfield, Blocton and Birmingham Railroad is not yet completed, so at present there are no facilities for shipping it from this basin. That part of this seam next to the South and North Alabama Railroad is so divided up by rival ownerships that there is little possibility of its being mined there until some of the owners either form a combination or solidify the tracts by purchase, thus making the tract area of fair working size.

The next seam in extent in this basin is the Nunnally seam, which the tests in this locality find to contain two feet nine inches of coal; still, a more thorough test along the outcrop may prove it to have a slightly larger or smaller average thickness. This basin has also a limited amount of the Wadsworth seam, with an average thickness of three feet three inches. This is a first-class seam for iron manufacturing purposes.

The following is a section of the Gould seam :

[*Gould seam in N. W. ¼ of N. W. ¼, in section 24, township 20 S., range 4 W*].

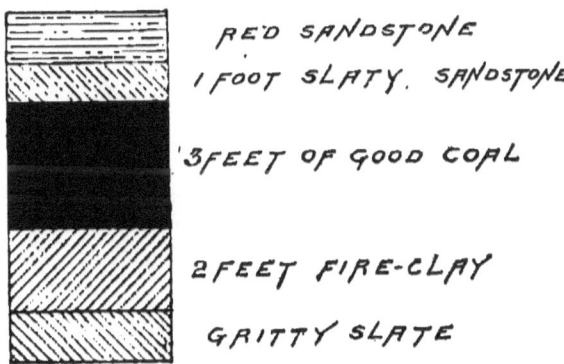

The *South and North Vertical Section*, and the *Dry Creek*

Horizontal Section from "K" to "L," on the accompanying map, will give the relative position of the seams in this basin; the Dry Creek Horizontal Section showing the form or structure of the basin and its connection with the Sub-Carboniferous and the Interior fault vertical measures.

The rate of dip of the measures of this basin varies mostly between fifteen and twenty-two degrees, and in some localities considerably more; the dip is nearly everywhere towards the southeast. There has been no mining hitherto in this basin as above stated, as it is only recently that railroads have begun to be constructed here. This, though, will soon be a thing of the past, for at present a great number of loud reports like the discharge of distant cannon can be heard daily and hourly made by the blasting operations going on in the construction of the Brierfield, Blocton and Birmingham Railroad through this basin.

NOTE.—I have the information from a source that appears to be trustworthy, that in S. 12, T. 20, R. 4 W, in Genery's Gap, the Brock seam has been exposed in the railroad cut, and shows a thickness of four feet. E. A. S.

CHAPTER IX.

THE LOLLEY BASIN.

The Lolley basin is situated to the east of Gurnee, to the southwest of Helena, and to the northwest of Montevallo; it is bounded on the north by the Piney Woods fault and Dry Creek basin, on the east by the great boundary fault, on the west by Dailey Creek basin and a portion of the Montevallo basin, on the south by the Montevallo basin and the anticlinal between it and the Lolley basin.

The following is a description of the boundary of the Lolley basin: Commencing at a point about two hundred yards northeast of Lacey depot on the Brierfield, Blocton and Birmingham Railroad; thence along the Piney Woods fault almost due west for about two miles along the fault thence along the said fault at a bearing of about S. 68° W., to the southwest corner of section 15, township 21, range 4 west; thence south and southeastwards up Jesse's Creek to the southeast corner of section 35, township 21, range 4 west; thence almost due east along the anticlinal between the Lolley and Montevallo basins to opposite Dogwood Grove Church on the east edge of the boundary fault; thence northwards along the west edge of the boundary fault, passing to the left of Mayline depot, continuing along the boundary fault to the point of commencement near Lacey depot.

This basin is drained by Piney Woods Creek, Beaver Dam Creek, Shoal Creek, King's Creek, Jesse's Creek, and Lick, or Big Creek.

The most prominent ridge in this basin is the "Divide," mostly called Pea Ridge, that separates the waters draining into the Cahaba river from those draining into Shoal Creek or Little Cahaba river; this divide commences west of the Mayline depot and southwest of the Lacey depot on the Brierfield, Blocton and Birmingham Railroad, and continues southwestwards dividing the drainage as aforesaid, down to where the Little Cahaba river joins the Big Cahaba river in Bibb county; this high and prominent ridge has been the great obstacle to the construction of a straight line of railroad through this part of the Cahaba Coal Field, the bend of the Brierfield, Blocton and Birmingham Railroad at Lacey depot became a necessity in order to obtain easy grades. This ridge is made by the Montevallo Conglomerate. The next prominent ridge is the one south of Piney Woods fault, commencing at the east edge of the coal field opposite William Lacey's farm and continuing westwards for four or five miles on the south side of Piney Woods Creek. There are also a number of irregularly formed ridges besides the above in other parts of the basin.

There are no public roads in this basin; what wagon roads there are in it, are better fitted for oxen than any other animals. The principal road in the basin is the one that leaves the Montevallo and Elyton road at William Lacey's and follows the top of the high ridge south of Piney Woods Creek, and leads on to the Henry Clark house; thence to the Anderson Allen house, here making a turn south and going to Newton Lolley's place, continuing on to the Bethel church on the Montevallo and Boothtown wagon road. The next wagon road in importance is the one leading from William Lacey place to Elias Walker's place, passing Dustin Dean's place and Isaac Walker's place on the way, then, at Elias Walker's branching off, one prong leading to Dogwood Station, the other to the Montevallo and Boothtown road at the Mrs. Lucas place, and to Bethel church by Newton Lolley's. These are all rough roads, and will not admit of hauling heavy loads along them. There are other roads to which the name of trails would be most appropriate, one going down Piney Woods Creek bank to the old Ryan place, another to the Henry Lee place, another to the Henry Lolley old place; these are partly grown up, and they are barely safe to venture along with a vehicle. The Elyton and Montevallo wagon road is a public road; it follows along the east boundary of this basin in 'Possum Valley but outside of the basin, passing close by

CAHABA COAL FIELD: LOLLEY BASIN.

Wilderness church, the Reneau place, Columbus Harper's, and the William Lacey farm.

The Brierfield, Blocton and Birmingham Railroad follows close along the eastern and northern boundaries of this basin, joining the Birmingham Mineral at Piney Woods Station and Gurnee Station, there connecting with Blocton and Bessemer and Birmingham, and the Birmingham Mineral Railroad to Helena and Birmingham; the south end of said road connects with the East Tennessee, Virginia and Georgia Railroad at a point one mile southwest of Montevallo.

The length of the Lolley basin is five and a quarter miles by an average width of three and fourteen hundredths miles; its surface area is sixteen and a half square miles. The amount of workable coal it contains, in seams of two feet and upwards in thickness, and within a vertical depth of 4,400 feet, is 357,000,000 tons (of 2,000 pounds). This computation makes no allowance for loss in pillars, or waste in mining.

The lowest workable seam outcropping in this basin is the Gholson; it outcrops in a few places along the Piney Woods fault, but in most places along this fault the seam is down in the fault. I have made a slight effort to cut its outcrop in that locality, but lack of time prevented me giving it a thorough test along the outcrop. This is an excellent seam with a good sandstone roof, in places having a thin layer of compact slate at the top of it; and it will average in thickness, in my estimation, four feet of good coal without slate partings. The next seam above this and outcropping farther south, is the Little Pittsburgh, then above this and underlying the Conglomerate, is the Thompson or Conglomerate seam, then still farther southward is the outcrop of the Helena seam, of which the following is a measured section:

[*Helena seam in section 18, township 21, range 3 W.; direction of strike N. 75° E.; S. 75° W. magnetic; direction of dip S. 15° E. magnetic; rate of dip 38°*].

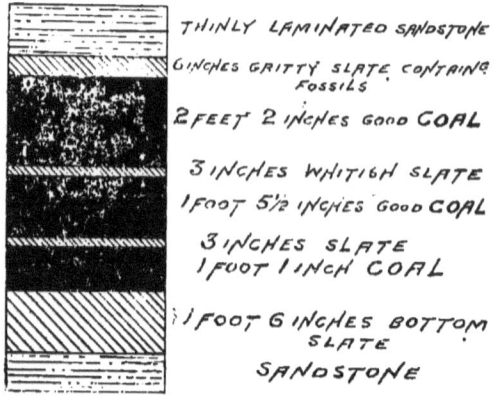

As will be seen the Helena has two thin layers of slate in it. The Helena seam has higher rate of dip here than it has further west, but is thicker at this point, having four feet eight inches of coal. The following is another section of the Helena seam with a less rate of dip:

[*Helena Seam in Section 13, Township 21 S., Range 4 W.; direction of strike N. 41° E.; direction of dip S. 49° E.; rate of dip 13°.*]

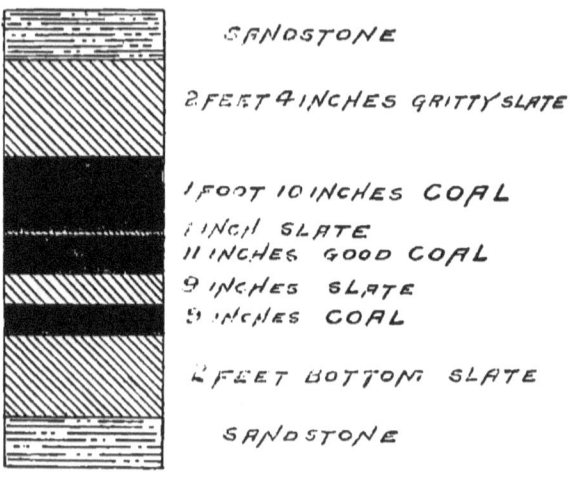

The next seam of workable thickness outcropping above this, and farther south is the Yeshic seam. While I have not seen this seam more than two and a half to three feet

CAHABA COAL FIELD : LOLLEY BASIN. 87

thick in this basin, yet in the Dailey Creek and Blocton basin it becomes four to five feet in thickness.

The next workable seam outcropping still farther south, in this basin is the Montevallo seam ; this seam is thicker here than its average thicknsss in the Montevallo Basin. The following is a measured section :

[*Montevallo Seam in section 24, township 21 S , range 4 W.*]

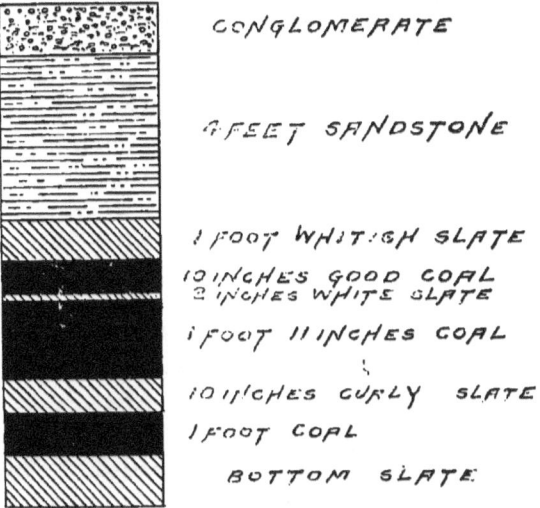

CONGLOMERATE

4 FEET SANDSTONE

1 FOOT WHITISH SLATE
10 INCHES GOOD COAL
2 INCHES WHITE SLATE
1 FOOT 11 INCHES COAL

10 INCHES CURLY SLATE
1 FOOT COAL

BOTTOM SLATE

Above the Montevallo there are nearly five hundred feet of conglomerate interlarded with sandstones and slate. In this conglomerate formation, there are four seams of coal, all of them either too thin or too impure to be workable. The first one, the "Air-shaft Seem," is about one hundred feet above the Montevallo ; the next one above this is the Black Fireclay seam of which the following is a measured section :

[*Black Fireclay Seam in the N. W. corner of section 35, township 21 S., range 4 W.; rate of dip 2°.*]

LAMINATED SANDSTONE
2 INCHES SOFT WHITE SLATE
5 INCHES SOFT COAL
½ INCH BLACK SLATE
1 FOOT COAL
½ INCH DARK COLORED SLATE
8 INCHES COAL
½ INCH WHITISH SLATE
4 INCHES COAL
BLACK FIRE-CLAY

The next seam above this is the Stine seam ; the top seam is the Luke seam, which can be seen above the Big Fall on Davis' Creek, at one of my test drifts made before or about the beginning of the late war ; the roof is a thick ledge of conglomerate.

A peculiar feature marks that part of the Cahaba Coal Field having the Montevallo seam underneath it, viz : the ground is covered with scattering pebbles where the conglomerate measures come to the surface ; where the sandstones outcrop an absence of the pebbles will be noticed through a belt or strip of country until the next ledge of conglomerate with its pebbles come to the surface. This is the case over a large area of the Lolley basin. The outcrop of the Montevallo seam on the accompanying map will show its limit.

There is another, and in places, a thick ledge of conglomerate over the Thompson seam ; it shows plainly on the surface, but this must not be confused with the conglomerate above the Montevallo, as it is a long distance underneath the Montevallo seam. There is another thin ledge of conglomerate still below the above, this one is near the lower bench of the Mammoth seam, or Clarke. This will not cause confusion in this basin as it is close to or in the Piney Woods fault.

The conglomerate formation above the Montevallo seam, has the purest springs of free stone water in the territory where they come to the surface, of any in this section of country. Wherever it forms the surface rock, its topography being high or rolling, it is remarkably healthy, probably more so than any other part of the State. For a more detailed statement or description of these ledges of conglomerate, see the section given in the first chapter. For the relative position of the seams of this basin, see the *Dailey Creek Vertical Section*, and the *Dry Creek Horizontal Section*, from "K" to "L," on the accompanying map.

The rate of dip of the measures of this basin varies from fifty degrees on its north edge, next to the Piney Woods fault, to one or two degrees at the synclinal south of the Elias Walker place. At a point at about half a mile east of the Elias Walker house Lick Creek falls about one hundred feet vertical over a perpendicular cliff of conglomerate; this is known in the settlement near as the "Big Falls." There has been no mining done hitherto in this basin; the country is sparsely settled, about two years ago six families were all the inhabitants it then had; they were Elias Walker and his son Isaac Walker, Newton Lolley, Anderson Allen, Henry Clark, and a well respected colored man named Dustin Lee and his family. The Lolley Basin is healthy but not well adapted for farming purposes, except along the creek bottoms.

My first examination of this basin was made in 1860, when I was employed by the Alabama Coal Mining Company to make a preliminary survey of their lands in this basin, and to make a more thorough survey of their lands in that portion of the Montevallo Basin which was then tapped by their branch railroad.

CHAPTER X.

THE MONTEVALLO BASIN.

The Montevallo Basin is situated to the northwest of Montevallo, and to the southeast of Gurnee. It is bounded on the north by the Lolley Basin, on the east by the great boundary fault that separates the Carboniferous from the Cambrian measures, on the southeast by the Overturned measures and the fault separating them from the Montevallo Basin, on the southwest and west by the Dailey Creek Basin, and on the north by the Lolley Basin.

The following is an outline of the boundary of the Montevallo Basin: Commencing at a point three hundred yards southeast of the Baker Mine entrance, at that part of the boundary fault where the fault immediately north of the "Over-turned measures" intersects it, thence south twenty-two degrees west, along the fault between the Over-turned measures and the Montevallo Basin a distance of one and three-quarter miles, to a point where that fault intersects Little Mayberry Creek; thence in a northwestwardly direction along the anticlinal, crossing Walker's Camp Branch, Jim's Branch, and Big Mayberry Creek, to the northwest corner of section 15, township 22, range 4 west; thence due north along the section line on the west side of sections 10 and 3, to the southwest corner of section 34, township 21, range 4 west; thence due northeast to the northeast corner of said section 34; thence southeastwardly up Jesse's creek to the southeast corner of section 35, township 21, range 4 west; thence nearly due east along the anticlinal between the Lolley Basin and the Montevallo Basin to nearly opposite Dogwood Grove Church at the east edge of the boundary fault, leaving the Davis Creek Falls to your right and the Ed. Davis' house to your left, to a point about three hundred yards southeast of Baker Mine, the point of commencement.

This basin is drained by King's Creek, Davis' Creek, Little Mayberry Creek, Walker's Camp Creek, Jim's Branch, Big Mayberry Creek, Lovelady Branch, Savage Creek, Rocky Branch and Jesse's Creek.

The highest and most prominent ridge in this basin is Pea Ridge (formed by the Montevallo conglomerate), a high ridge, flat in places, that divides the waters draining into Little Cahaba River, and those draining into the Big Cahaba River; it is irregular in shape, becoming high between the head waters of the creeks and branches that drain it. Its altitude above Shoal Creek is over 400 feet in places. There are various other ridges also due to the Montevallo conglomerate, between the head waters of Big Mayberry Creek, Jim's Creek, Little Mayberry Creek and Davis' Creek that are in vertical height above Shoal Creek over three hundred feet of barometrical measurement. The remarkable feature of these ridges, is the immense amount of conglomerate pebbles scattered over the ground, where the different layers of the great Montevallo conglomerate (above seam of same name) crop out at the surface; all of the high lands underlaid by this Montevallo conglomerate are remarkably healthy.

The principal wagon roads of this basin are the Montevallo and Boothtown or Gurnee road; the Columbiana and Booth's Ferry road; the Aldrich and Blocton road; the road from Bethel Church along Pea Ridge; and the Aldrich and Dogwood Grove road; besides these there are various other roads partly grown up with undergrowth, and former roads that are now used as cattle trails or bridle paths.

Of railroads in this basin, the Brierfield, Blocton, and Birmingham railroad runs close along its eastern edge, with stations at Dogwood and at Aldrich; the Montevallo Coal and Transportation company have a short line of railroad running from their slope in the Montevallo seam, in the southeast quarter of section 24, township 22, range 4 west, and joining the Brierfield, Blocton and Birmingham railroad a short distance south of Aldrich depot; these are all the railroads connected with the basin at present.

This basin is four and one-tenth miles (4 1-10) in length,

by an average width of three and three tenths (3 3-10) miles, and contains a surface area of thirteen and eighty-six hundredths (13 86-100 square miles.

The amount of workable coal it contains in seams of two feet and upward in thickness, is 300,000,000 of tons (of 2,000 pounds,) without any allowance being made for loss in pillars or waste in mining.

The lowest workable seam outcropping in this basin is the Montevallo seam; it is also the highest outcropping workable seam in the basin. There are six other seams outcropping in this basin besides the Montevallo seam, two below and four above the Montevallo, but all six are either too thin or too impure to be workable. My examinations and tests of the most of these thin seams were made twenty-eight years ago; I have tested the others at various times since. My tests in the two below the Montevallo were made on Walker's Camp Branch; the Air Shaft seam was tested near the Baker mine; the Black Fireclay seam test is on the headwaters of Jesse's Creek; my tests on the Stine seam and the Luke seam were made on Davis' Creek; the only workable seam discovered yet, outcropping in the Montevallo basin is the Montevallo seam; this seam was discovered and mined three or four years before the beginning of the war.

The writer mined this seam on a lease from the Alabama Coal Mining Company and Montevallo Coal Company in 1859, shipping by what is now known as the East Tennessee, Virginia and Georgia Railroad to Talladega and Selma, thence by Alabama river to Montgomery and Mobile. It was then considered the best domestic coal mined in the State. In fact, up to January, 1860, it was the only coal in the State that was shipped to market by railroad. The average thickness of this seam is from two and a half feet to two feet nine inches. The following is a section of it:

CAHABA COAL FIELD: MONTEVALLO BASIN.

[*Montevallo seam in S. E. ¼ of S. W. ¼, of section 24, township 22 S., range 4 W*].

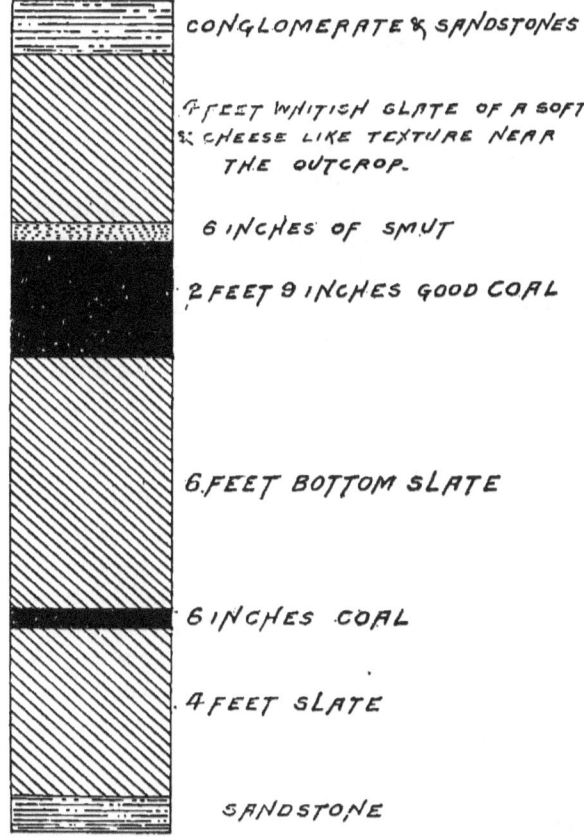

The method of mining it is, first use a light mining pick and pick out the whole or part of the smut above the coal, then blast the coal out with powder or wedge it up with hammer and wedges. When blasted without first using the pick, the coal is more shattered and the amount of slack is increased.

For relative positions of the seams in this basin see the *General Vertical Section* and *Montevallo and Blocton Horizontal Section* from "M" to "N" on accompanying map.

It will be seen by these sections that all the other workable seams of the Cahaba Coal Field are in this basin and underneath the Montevallo seam, so that the portion of this basin that has the Montevallo seam under its surface, contains all the workable seams of the Cahaba Coal Field.

The rate of dip of the measures of this basin, varies from 9° to flat or level measures in the synclinal part of the basin; a large area along the synclinal of this basin is perfectly level.

For a distance of about two miles west and northwest of Aldrich depot on the Brierfield, Blocton and Birmingham Railroad, the Montevallo seam has been worked by various companies in the past thirty-four years; at present the only parties engaged in mining it are the Montevallo Coal and Transportation Company, of which William F. Aldrich is president, and James L. McConaughy, secretary and treasurer. They have a good mile opened on the seam by slope, and are well able to supply the present demand for Montevallo coal.

The 500 feet of measures above the Montevallo seam are a series of conglomerate ledges interlarded with pebbly sandstones and with sandstones. About the middle of these measures there is a fifty feet layer of dense conglomerate; this forms several "falls" on the creeks and branches of the Montevallo and Lolley basins; the four thin seams "Air Shaft," "Black Fireclay," "Stine," and "Luke" are imbedded in the above mentioned 500 feet of measures.

The layers of conglomerate vary in thickness and position; the plate next above the Montevallo seam is at places close down on the seam, while at other places it is 35 to 40 feet above it.

Analysis of coal from the Montevallo seam, from Montevallo Coal and Transportation Company's slope, Aldrich, Ala., by J. L. Buson:

Moisture	1.858
Volatile matter	36.592
Fixed carbon	54.002 } Coke......... 61.550
Ash	7.548
	100.000
Sulphur in coal	1.726
Sulphur left in coke	1.156
Per cent. of sulphur in coke	1.878

CHAPTER XI.

THE OVERTURNED MEASURES.

The Overturned Measures are situated to the west of Montevallo and to the northwest of Brierfield depot and rolling mills.

The Overturned Measures are bounded on the north by the fault that separates them from the Montevallo basin and Dailey Creek basin; on the east by the great boundary fault that separates the Carboniferous and Cambrian measures; on the south by the same great boundary fault that follows along the south edge of the Cahaba Coal Field.

The following is a rough outline of the boundary of the Overturned Measures: Commencing at the great boundary fault about three hundred yards southeast of the Baker mine entrance; thence southeastward along the fault that separates the Overturned Measures from the measures of the Montevallo and Dailey Creek basins, about two and a half miles; thence along the fault nearly due west about three and a half miles to the middle of section 5, township 24, range 11 east; thence southwestward along said fault to the half mile post on the south side of section 15, township 24, range 10 east, (this point is at the south boundary of the coal field;) thence eastwardly and northeastwardly along the boundary fault to the southwest corner of section 5, township 24, range 12 east, (this point is nearly opposite Thompson's mill on Shoal Creek;) thence along the boundary fault nearly due north, to the point of commencement, three hundred yards southeast of the Baker mine entrance.

The Overturned Measures are drained by branches running into Shoal Creek; by Little Mayberry Creek, Big Mayberry Creek, east prong of Four Mile Creek, west prong of Four Mile Creek, Alligator Creek, and some small branches running into Little Cahaba river.

The most prominent ridge in the Overturned Measures is the Conglomerate ridge, immediately south of and parallel with the fault that separates the Overturned Measures from the Montevallo and the Dailey Creek basins ; there are other ridges of lesser prominence between the outcrops of the seams and following parallel with them, but they are not so continuous as the Conglomerate ridge near the north edge of the Overturned Measures.

The principal wagon roads in the Overturned Measures are as follows : the road leading from Montevallo to the old shaft ; the road leading from the Irish Pit to Thompson's Mill; the road leading from the Irish Pit to Peter's Mines ; the road leading from Pea Ridge to Potts' Tan Yard and to Peter's Mines ; the road leading from the Rainey slope to Montevallo ; the road leading from Berea Church to the Brierfield Coal and Iron Company's Smelting Furnace.

Of railroads in the Overturned Measures the Brierfield Coal and Iron Company's Branch Railroad runs through a portion, connecting the company's coal mines, (known in the neighborhood as Peter's Mines), with the East Tennessee, Virginia and Georgia railroad ; the Brierfield, Blocton, and Birmingham railroad runs along the east end of the Overturned Measures ; the branch railroad of the Montevallo Coal and Transportation company also runs through a portion of the East end of the Overturned Measures, connecting their slope in the Montevallo seam with the Brierfield, Blocton, and Birmingham Railroad.

Twenty-nine years ago a branch railroad extending from what is now called Birmingham Junction Depot, out to the "old office," and from there was connected by tram-road with the "old shaft" or slope in one of the Overturned seams. The tram-road and a portion of said branch railroad are now abandoned.

The Overturned Measures are ten and a quarter ($10\frac{1}{4}$) miles in length by an average width of about one mile ; the surface area is ten and a quarter square miles.

The amount of workable coal in seams of two feet and upwards in thickness in the Overturned Measures, is 167,000,000 of tons (of 2,000 pounds) with a vertical depth of 4,500 feet.

The conglomerate and the seams outcropping immediately south of it, viz : the Dodd seam, Cooper, Shaft, Beebee and the Cannel seam are all overturned; they all outcrop on Little Mayberry Creek and on the Big Mayberry Creek. The four hundred feet of conglomerate and sandstones next the fault, forming the north boundary of the Overturned Measures, is a part of the top or cap rock of our Alabama Coal Measures; this is the lower part of the great Montevallo conglomerate. In examining all the above mentioned seams, the bottom slate was found to be on top in every case. a.

The angle or rate of dip of these seams, varies from fifty-six degrees at the Cannel seam, to sixty degrees at the Cooper seam. I have examined these measures closely along their outcrops for over seven miles, and find them overturned the whole of that distance. The best point for examination of this portion of the Overturned Measures, is on Little Mayberry Creek about five hundred yards west of the old shaft or slope. The old shaft or slope was worked by the Montevallo Coal Mining company twenty-nine years ago, under my superintendence; I had then an excellent opportunity to obtain a thorough knowledge of that part of the Overturned Measures.

The Little Mayberry Creek at this point cuts in a direct course through the steep dipping measures that contain the above mentioned seams. The relative position of these seams is as follows: Commencing at the "fault" on Little Mayberry Creek, where you can put one foot on the Overturned Measures, dipping at a rate of sixty degrees, and the other foot on the flat measures of the Montevallo Basin dipping only two or three degrees; thence southward down the creek, passing various ledges of conglomerate interlarded with sandstones on the way, a distance along the surface of three hundred and ninety feet (390); you have now passed over three hundred and thirty-eight (338) feet in thickness of measures. This brings you to the Dodd seam, and you have just passed over three hundred and thirty-eight feet of the lower part of the Montevallo conglomerate. The Dodd vein is the Montevallo seam. Continuing down the Little Mayberry Creek seventy-three feet

---

a See Chapter I, and Introductory Chapter for further mention of the reversal of the strata.

farther, passing over sixty-three feet in thickness of measures, you arrive at |the Cooper seam, which is the under seam of the Montevallo. (This underseam is exposed in the Dailey Creek Basin at a point three miles northwest of where it intersects Little Mayberry). Continuing on down the creek a distance of three hundred and twenty feet, the rate of dip being sixty degrees all the way from the "fault," you pass over since leaving the Cooper, two hundred and eighty (280) feet in thickness of measures, and have arrived at the Helena seam, of which the following is a section.

[*Helena Seam in section 1, township 24 N., range 11 E. Rate of dip 65°*]

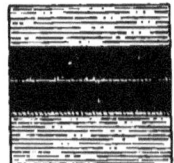

Continuing on down the creek one hundred and forty-two (142) feet farther, passing over one hundred and twenty-four (124) feet in thickness of measures, you arrive at a ledge of conglomerate, (the previous four hundred and sixty-seven (467) feet in thickness being nearly all sandstone); thence down the creek a distance of two hundred and twenty-five (225) feet, passing over one hundred and ninety-seven (197) feet in thickness of measures, you arrive at the Shaft seam, of which the following is a section.

[*Shaft Seam in section 1, township 24 N., range 11 E. Rate of dip 65°*]

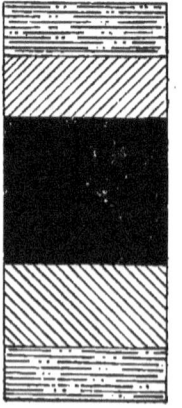

Continuing on down the creek seventy-three feet farther, passing over sixty-three feet in thickness of measures, you arrive at the "Three Feet Seam;" continuing on down the creek, a distance of three hundred and thirty-six (336) feet, you pass over two hundred and eighty-eight (288) feet in thickness of measures, and arrive at the Beebee seam; thence down the said Little Mayberry Creek, square across the measures a distance of five hundred and twenty-nine (529) feet, passing over four hundred thirty-eight feet in thickness of measures, you arrive at the Cannel seam. The rate of dip of the rocks you have passed over are as follows: at the conglomerate between the "fault" and the Dodd seam the rate of dip is sixty-one degrees; at the Helena, sixty-one degrees; at the Shaft seam, sixty degrees; at the Beebee seam, fifty-nine degrees; and at the Cannel seam, fifty-six degrees.

The average thickness of the above mentioned seams, as evidenced by the tests made, are as follows:

Dodd, 4 to 6 feet.
Cooper, 2¼ feet.
Shaft seam, 4 feet.
Three Feet, 2½ to 3 feet.
Beebee, 3 feet.
Cannell, 3 feet, part of it bony.

For relative position of the seams of the Overturned Measures, see the *Little Mayberry Creek Vertical Section* on the accompanying map. The seams near the south boundary of the Overturned Measures have been worked for several years by the Brierfield Coal and Iron Company at what is known as Peter's mines; these seams have a south or southeast direction of dip, the same as the Dodd, Shaft, Beebee, and Cannel seams, on Little Mayberry Creek.

The company sunk two slopes on the Lemley or B. seam, and from the bottom of this slope they tunnelled to the C. or "Cubical vein," and to the D. or "Figh seam;" they also tunnelled southwards to the A. seam, and hoisting the coal from all of them at the B. slope in the Lemley seam. My examination of these seams was made in 1859, when I gave to the B. seam the name of "Lemley," part of it being then owned by an old planter named Mennis Lemley, living on the plantation just south of it; I gave the C. seam the name of "Cubical seam" on account of its having a cubical fracture; and named the D. seam "Figh seam," in remembrance of my friend George M. Figh, who died in Dallas, Texas.

In passing through by Peter's mines slope in April, 1890, I noticed that the B. slope was stopped.

I do not remember whether my examination of these seams in 1859 decided the question as to whether they were overturned; that is, the bottom slate on top like the Dodd, Cooper, Shaft, Beebee, and Cannel seams, or not.*a*

There is a thin seam between the B. and C. seams of about two to two and a half feet in thickness, that has never been worked. At the boundary fault, south of Peter's mines, there is an outcrop on Shoal Creek in section 12, township 24, range 11 east, that bends over and forms a complete arch, plainly to be seen exposed on the bank of the creek thirty-one years ago; it may be covered up now by the falling in of the creek bank. This is one of the seams of the boundary fault measures. If the Figh, Cubical, Lemley, and A. seams are not overturned with the bottom slate on top like the Dodd, Cooper, Shaft, Beebee, and Cannell seams on Little Mayberry Creek, then there must be a fault between the two series of seams. I have not seen any surface evidence of any fault between them, more than the "hitch" in the measures about the middle of section 12, forming a slight zig-zag in their outcrops.

The first mining done in the "Overturned Measures" was by the Alabama Coal Mining Company in or about the year 1857, when they opened a series of "drifts" on Little Mayberry Creek, in the Cooper seam, the Shaft seam, and Beebee seam; then in the year 1859, the company sunk a slope on the Shaft seam to a depth of about 160 feet along the slope, the seam having a rate of dip of 60° to 61°. The company obtained a hoisting engine and boilers from Wilkesbarre, Pennsylvania, the cylinder of which is now in the scrap pile at the Shelby Rolling Mill, Helena. About

---
*a*It seems most probable that these seams also are overturned, for at Thompson's Mill, a quarter of a mile south of the Lemley seam, occurs the instance of a coal seam with Cambrian rocks immediately above it, shown in the illustration given in the introductory chapter. E. A. S.

this time the company acquired some new stockholders and changed the name of the firm from Alabama Coal Mining Company to Montevallo Coal Mining Company, but I do not remember the exact date of the change.

The company found it necessary to bring men from Pennsylvania to fit up the engine and hoisting machinery; one of them, John Hartley, an Englishman, was brought to build the engine bed and boiler masonry.*b*

Some machinists also came at the same time Hartley did.

The company had gotten the slope sunk by means of horse power to the depth of 160 or 165 feet, and had driven the gangways out one or two hundred feet previous to my taking charge as superintendent of the company's works, obligating myself to keep the underground surveys advanced up to the full progress of the work at the end of each month, and furnish the company with a geological map showing the seams on their property, which was done under some difficulties.*c*

The aforesaid hoisting engine, boilers, and machinery from Wilkesbarre, Pennsylvania, was the first steam power machinery for hoisting coal ever used in Alabama.

The stockholders of the company who first commenced to use the aforesaid hoisting machinery, were Col. John S. Storrs, of Montevallo, president of the company; Judge Cooper, of Lowndes county; Dr. Miller, of Wilcox county; Alexander White, of Selma and Talladega; Gen. C. Robinson, of Lowndes or Wilcox counties, and John R. Keenan, of Selma, Ala., etc. These were the principal stockholders when the machinery was obtained. A little later on ex-Gov. T. H. Watts, George M. Figh, Benjamin B. Davis, and Dr. I. T. Tichenor, all of Montgomery, became stockholders in the Montevallo Coal Mining Company, so it will be seen

---

*b*Hartley, soon after his arrival, told me he had been advised to bring a bowie knife and carry it with him all the time he was here; after enjoying a good laugh at his expense for his causeless fears, I advised him to keep away from bar rooms and grog shops, and bury that knife until he started back to Pennsylvania.

*c*My first map presented to the board of directors showing the outcrop of the Montevallo seam, near where the mining is now going on, as shown on the accompanying map, was made on strong brown paper, called cotton paper, as it was mostly used to wrap up cotton samples in.

that the first efforts at the scientific mining of coal with steam machinery in Alabama were made by men mostly from the "Black belt" portion of the State.

Analysis of Coal from "B." Seam of the Brierfield, Bibb County, Ala., by J. L. Beeson.

Moisture	2.265
Volatile matter	57.130
Fixed carbon	37.407 } Coke......... 40.605
Ash	3.198
	100.000
Sulphur in coal	1.158
Sulphur left in coke	.487
Per cent. of sulphur in coke	1.198

CHAPTER XII.

THE DAILEY CREEK BASIN.

The Dailey Creek basin is situated to the east and northeast of Blocton, to the west and northwest of Montevallo, and to the southwest of Helena, Gurnee being in the north end of this basin. It is bounded on the northwest by the "Interior fault" and the Blocton basin, also by a portion of the Gould basin ; on the north and northeast by Dry Creek basin and Lolley basin, on the east side by the Montevallo basin, and on the south side by the "Overturned Measures" and the "South boundary fault."

The following is a description of the boundary of the Dailey Creek basin : Commencing at the gap in the Conglomerate ridge where the Little Mayberry Creek cuts through it, at the fault where the flat measures and the "Overturned" measures come close together, thence northwestwardly along the anticlinal to the northwest corner of section 15, township 22, range 4 west, thence due north along the section lines on the west side of sections 10 and 3, to the southwest corner of section 34, township 21, range 4 west; thence northeast to the northeast corner of said section 34; thence northwestward down Jesse's Creek to the southwest corner of section 15, township 21, range 4 west; thence northwest to the southeast edge of the Interior fault vertical rocks near the northwest corner of section 16, township 21, range 4 west ; thence southwestward along the southeast edge of the Interior fault leaving Boothtown to your left; thence close by Cadle Station, crossing the railroad at this point, close by the Gardner old mine ; continuing close along the edge of the Interior fault to the edge of the coal field at a point about a quarter of a mile west of the southeast corner of section 17, township 24, range 10 east; thence eastward along the boundary fault; after advancing two hundred yards you will pass close by the left side of the Joseph Lightsey house; continuing along the boundary fault, crossing Cahaba river about two hundred yards above the "boat landing" to the half mile post on the south side of section 15, township 24, range 10 east; thence northeastwards to the middle of section 5, township 24, range 11 east; thence eastwardly along the line of fault forming the north boundary of the "Overturned measures" to the Little Mayberry Creek, at a point about 700 yards to the northwest of the old Shaft seam slope, this being where the rocks of the Montevallo basin and the Overturned measures come together, the point of commencement.

The Dailey Creek basin is drained by the Cahaba river and its tributaries: Jesse's Creek, Rocky Branch, Lick Creek, Savage Creek, Lovelady Branch, Glade Branch, Hudgin's Creek, Swep Branch, Thrasher's Field Branch, Stone Coal Branch, Dailey Creek, Short Creek, Big Lick Creek, Beech Camp Branch, Pine Island Branch, Big Ugly Creek, Little Ugly Creek, Four Mile Creek, and Alligator Creek, the last two emptying into Little Cahaba River, all the others drain into the Big Cahaba river.

The most prominent ridge in this basin is Pea ridge, and its continuation southwest, forming the "divide" between the waters of Little Cahaba river and the Cahaba river. This "divide" forms a broad, high ridge for a length of about nine miles in this basin; its full length is much more, as it continues northeast nearly to Lacey Station, at the head of Piney Woods Creek. Its full extent is from near Lacey Station to the forks of the Big Cahaba and Little Cahaba rivers. On the northwest side of this ridge the waters drain into Big Cahaba river, and on the southeast side the waters all drain into the Little Cahaba river. This ridge or "divide" has an altitude in places of 400 feet above the river.

The next most prominent ridge is formed of the roof rock of the Gholson seam. The roofs of the Coke seam and the Thompson seam both form high ridges in portions of this basin.

Of the wagon roads of this basin the principal one is the Montevallo and Tuscaloosa, or Booth's Ferry road; this is a county road, on which vehicles can be used. Another

CAHABA COAL FIELD : DAILEY CREEK BASIN. 105

wagon road leads from the Aldrich mines near Montevallo to Blocton, going by Berea church and crossing the river at Lily Shoals. Another wagon road leads from Berea church to Potts' Tan yard. Another wagon road leads from Peter's mines to the James Rich ford on Cahaba river.

Two railroads enter this basin at its north end, the two uniting near Gurnee or between Gurnee Station and Piney Woods Station; one of the railroads is the Birmingham Mineral Railroad, extending from the Louisville and Nashville Company's main line at Helena, to its junction with the Brierfield, Blocton and Birmingham Railroad, near Gurnee. The other road is the Brierfield, Blocton and Birmingham Railroad which extends from Birmingham Junction Station near Montevallo, to Gurnee and Blocton. These two railroads have been recently constructed and are both now completed and in running order.

The Birmingham Mineral Railroad Company have a lease from the Brierfield, Blocton and Birmingham Railroad Company, enabling them to run their trains clear through from Helena to Blocton.

The Brierfield, Blocton and Birmingham Railroad Company are now building a railroad from Gurnee to Bessemer and Birmingham; the whole line being now constructed under contract let to Aldrich, Worthington & Co., railroad contractors.

Two years ago, and prior to the construction of these railroads, the Dailey Creek basin did not have a population of more than an average of one family to the square mile, but since that, the Excelsior Coal Company have opened their two new slopes, and miners with their families have gone to live near the mines. The population has thus increased to ten times what it was two years ago.

The Dailey Creek Basin has a length of thirteen miles by an average width of three and two-tenths miles, and contains a surface area of forty-one and a half square miles ; it contains of good workable coal in seams of over two feet in thickness, and within forty-five hundred feet in vertical depth seven hundred and seventy-one millions of tons, (771,000,000—of 2000 lbs.) In computing this estimate of amount of coal in the basin I have made no allowance for loss in pillars, or waste in mining.

The lowest workable seam outcropping in this basin is the seam known as the "Big Vein." This seam is the Wadsworth of the South and North Alabama railroad. Near Boothtown it runs into the vertical measures of the "Interior Fault." Its thickness in the south end of the basin is eight feet in the aggregate; a part of this, though, is impure and shaly, but probably four feet of good coal can be gotten out of it.

The most workable seam is the "Clean Coal Seam," which is only two and a half feet in thickness. The next workable seam above this is the "Beech Tree seam," of three feet in thickness of good coal; the "Half Yard coal" comes in between the two last mentioned seams. A short distance above the Beech Tree seam is a thin seam of six inches; this, with the "Clean Coal," "Half Yard" and "Beech Tree," forming a group of four seams between the Big Vein and Coke seam. Between this group and the Coke seam, is a thin seam that becomes sixteen inches thick in places. Then above this is the Coke seam. This seam near Dailey Creek, ranges from three to three and a half feet in thickness, and is a good coal, making an excellent coke. There are two thin seams a few inches thick above the Coke seam, but the next workable seam is the Clark seam, which, when discovered thirty years ago, was named the "Spring vein." The Clark varies in size from two and a half to four feet in thickness, and is of very good quality. Above the Clark, a varying distance of from ten to a hundred feet is the Gholson seam; this is a remarkably good seam of solid coal, varying from four to five feet in thickness with a good sandstone roof. From my remembrance of measurements made in the old Gholson mine twenty-five years ago, when the mine was still open, the average thickness of the seam through the mine was five feet. When the Gurnee workings have advanced to flat part of the basin, the Excelsior company will have an excellent seam, with a good roof and an immense area of flat or level measures to work in. The following are measured sections of the Clark and Gholson seams:

[*Clark Seam in section 16, township 21 S., range 4 W. Rate of dip 16°*]

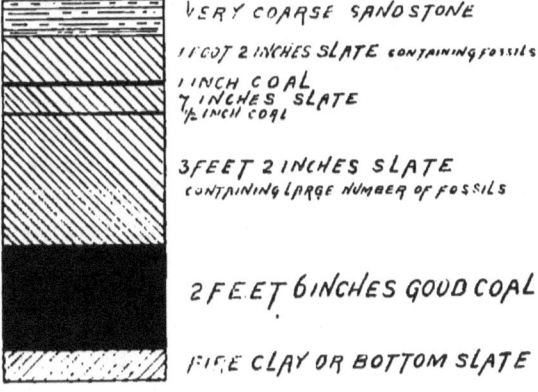

[*Gholson Seam in section 21, township 21 S. range 4 W. Rate of dip 16°*]

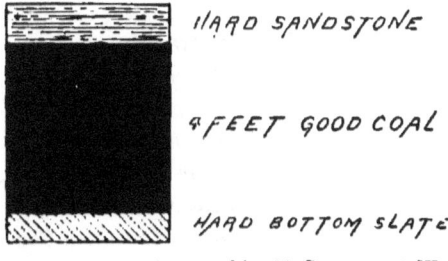

[*Gholson Seam in section 12, township 22 S., range 5 W. Direction of Strike, N. 34° E. Direction of dip, 56° E. Rate of dip 9°*]

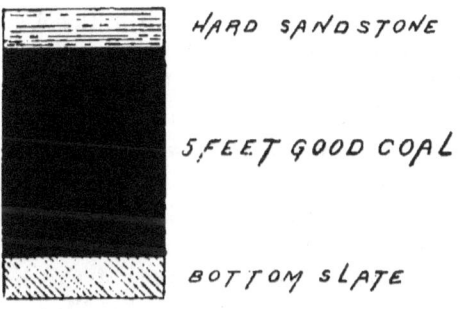

The next seam of workable size is the Middle Vein, of two and a half feet in thickness. This is the "Little Pittsburg Seam" of the South and North Alabama railroad. Above the "Middle Vein" are two thin seams, representing the "Quarry seam" and the "Smithshop seam" of the South and North Alabama railroad company. Above these is the

Thompson or Conglomerate seam, varying in size in this basin, from three to six feet. A short distance above this seam is a ledge of conglomerate that is fifty feet thick in places, but in other places, only a few feet. The next seam above this is the Helena; this seam in this basin varies in size from one and a half to four feet, and in some places is divided up into two or three benches, with slates intervening. The next seam above this is the Yeshic seam; a seam that is generally four to five feet in thickness; its condition is mostly impure in this basin. The next workable seam above this is the Montevallo seam of two and a half to four feet in thickness. For sections of this seam, see the chapters on the Lolley Basin and Montevallo Basin. This seam has about the best reputation for a good domestic coal, of any in the State. The outcrop of it can be seen beneath a ledge of conglomerate on a branch, a few hundred yards south of Antioch Church; the branch empties into Savage creek. The four thin seams above the Montevallo seam are the "Air Shaft seam," "Black Fireclay seam," "Stine seam," and the "Luke seam;" none of them are workable, and they vary so in thickness and amount of impurities, that they are not worth the reader's attention, though a section of the "Black Fireclay seam" can be found in the chapter describing the Lolley Basin.

The measures of the north end of the Dailey Creek Basin, dip towards, and are connected with the Lolley and the Montevallo Basins. The largest and most important of the seams of the Lolley and Montevallo Basins can be worked by slopes driven down from their outcrops in the Dailey Creek Basin. The anticlinal between the Lolley and Montevallo Basins appears to be pointing in the direction of Jesse's Creek; the lower rate of dip than usual in the lower part of Jesse's Creek is probably due to the said anticlinal.

For relative position of the seams of this basin, see the *Dailey Creek Vertical Section*, and the *Blocton and Montevallo Horizontal Section* from "M." to "N." on the accompanying map.

The rate of dip of the measures in this basin, varies from forty-five degrees at the Big Vein, to ten or fifteen at the Gholson seam, down to one or two degrees or flat, at the synclinal east of Berea Church ; most of the southeast side of the basin is flat or nearly flat.

The first mining done in this basin was during the war between the States, by refugees from Mississippi and elsewhere. They were Brooks and Gainer, mining close to where Gurnee now is. Rogers; Carter; Gholson & Co.; Herndon, and Thompson. They hauled their coal in wagons to the nearest point on the Selma, Rome and Dalton Railroad. The coal was used by the Confederate Government at the arsenal at Selma. The seams worked by them were the Clark seam, the Gholson seam and the Thompson seam. These three seams were all they mined in this basin ; their method of mining was by "drift," and horse power slopes ; none of them used steam power in any shape. The distance from their mines to the railroad was by the wagon road about twelve miles, and with a team of four mules and wagon, they hauled a ton per day to the railroad per each team ; this was counted a day's hauling.

None of them advanced their mine workings very far from the outcrop, their principal work being hauling the coal and keeping their long wagon roads in hauling condition.

All of these mines stopped when the war ended ; the refugees then, with one or two exceptions, went back to their former homes. Since that time the mines have been abandoned and grown up with briars, till about January, 1889. From this date railroads have been built, connecting this region with Montevallo and Selma, Blocton, Bessemer and Birmingham, and with Helena, Montgomery and the Gulf, and, by means of the steam colliers now running from Pensacola, with Havana and all the coal markets in the Gulf of Mexico.

The contrast between the appliances and methods of mining used in the basin twenty-five years ago, and those used at present, is very great.

Since January, 1889, the Excelsior Coal Company have sunk two large slopes on the Gholson seam ; one of them, No. 1, or Gurnee Slope, is now down eight hundred feet ; these slopes, if continued on in the direction they are now being driven, will penetrate an immense region of flat, or nearly level seams, sufficient to furnish continuous work for several generations of miners.

Analysis of coal from the Gholson Seam, Slope No. 1, Gurnee, Alabama, by J. L. Beeson.

Moisture 1.589
Volatile matter 35 760
Fixed Carbon .... 58.871 } Coke..62 651.
Ash. 3.780

100.000

Sulphur in coal 1.547
Sulphur left in coke .... .781

Percentage of sulphur in coke 1.249

CHAPTER XIII.

THE BLOCTON BASIN.

The Blocton basin is situated to the south and southwest of Bessemer, to the southeast of Woodstock and Vance's, to the north of Centreville, to the west of Aldrich and Montevallo, and to the southwest of Gurnee, Blocton occupying the middle portion of the basin.

This basin is bounded on the north by the Gould basin, on the northwest by the Sub-Carboniferous measures, at the visible portion of the southwest end it is bounded by a large deposit of "Drift measures" overlying and completely hiding the Carboniferous from sight, on the south it is bounded by the great boundary fault, and on the southeast side it is bounded by the Interior fault vertical coal measures, beyond which is the Dailey Creek basin.

The following is a description of the boundary of the Blocton basin: Commencing at the northwest edge of the Interior fault opposite Booth's Ferry in the south half of section 19, township 21, range 4 west; thence northwest along the Booth's Ferry and Tannehill wagon road, to the sharp bend in Sand Mountain in the south half of section 3, township 21, range 5 west; thence northwest along the base of the Millstone Grit nearly one mile, to where Sand Mountain makes another sharp turn; thence southwestward along the base of the Millstone Grit of Sand Mountain; the red fossiliferous ore cropping out about half a mile to the right. Then crossing the Cahaba Coal Mining Company's Railroad at "Thrasher's Mill," on the township line, between townships 21 and 22, and continuing along the base of the Millstone Grit, crossing Hill's Creek about three-quarters of a mile northwest of Randolph's Mill, and crossing Schultz's Creek at Burt's Mill; thence along the base of the Millstone Grit to the half mile post on the south side of section 22, township 24, range 8 east. To the southwest of this the Carboniferous is completely covered with drift. Thence southeast to the half mile post on the west side of section 6, township 23, range 9 east; thence northeastwards along the boundary fault, crossing Schultz's Creek about a quarter of a mile north of the wagon road bridge; passing Schultz's Creek church about 700 yards to the north of it, and continuing on along the boundary fault to a point two hundred yards west of Joseph Lightsey's house in the northeast quarter of the northeast quarter of section 20, township 24, range 10 east; thence northeastwards along the northwest side of the vertical measures of the "Interior fault," crossing the railroad about half a mile southwest of Cadle Station, and crossing the Cahaba river near the half mile post at the south side of section 2, township 22, range 5 west; continuing northeastwards along the northwest edge of the Interior fault vertical measures to opposite Booth's Ferry, in the south half of section 19, township 21, range 4 west, the point of commencement.

The Blocton basin is drained by the Cahaba river and its tributaries, Shades Creek, Cane Creek, Little Cane Creek, Bear Branch, Big Ugly Creek, Little Ugly Creek, Caffey's Creek, Turkeycock Branch, Lick Branch, Green Branch, Pratt's Creek, Stone Quarry Branch, Hill's Creek, Schultz's Creek, and Haysop Creek, the waters of all these creeks and branches finally reach Cahaba river. It is along the valley of one of these creeks (Caffey's Creek) that the Cahaba Coal Mining Company built their railroad, enabling them to open up their mines in this basin; this was the easiest route by which they could get railroad access to the seams in this basin, though the engineering difficulties of the route brought the cost of their nine miles of railroad up to over $160,000.

The most prominent ridges of this basin are Sand Mountain, formed of the lower portion of the Millstone Grit, extending all along the northwest side of the basin, though it is a little broken at its southwest end. The next ridge in prominence is the ridge formed of the roof rock of the Underwood or Thompson seam.

This basin, like all other parts of the Cahaba Coal Field, is not well provided with good wagon roads. The principal ones in the basin are the Woodstock and Blocton road, the Blocton and Pratt's Ferry road, (this is what the settlers designate as the new cut,) the Blocton and Centreville road, the Blocton and Gurnee road, the Woodstock and Centreville road, the Tuscaloosa and Pratt's Ferry road, Boothtown and Greenpond road, Blocton and Shades Creek church or Helena road, and the Scottsville and River Bend road.

The railroads in this basin are the Cahaba Coal Mining Company's Railroad, connecting their Blocton mines with the Alabama Great Southern Railroad at Woodstock, and with the Blue Creek extension of the Birmingham Mineral Railroad at the Blocton Junction depot near Woodstock.

There is another railroad recently completed that enters the basin from the east side, coming from Montevallo to Gurnee, and from Gurnee to Blocton, constructed by the Brierfield, Blocton and Birmingham Railroad Company over the Gurnee and Blocton portion of which the Birmingham Mineral Company have a lease or right to run their trains to Blocton, from their Helena and Gurnee branch.

This gives the Blocton basin connection with the Alabama Great Southern Railroad, the Birmingham Mineral system, and Louisville and Nashville Company's main line, and the East Tennessee, Virginia and Georgia main line by means of the Selma, Rome and Dalton Division, which are three of the most important mineral railroads in the State.

The Blocton basin is eighteen miles in length by an average width of five and a quarter miles. Its surface area is ninety-four and a half square miles, and it contains, in seams of workable coal of two feet and upwards in thickness, and within 3,800 feet of vertical depth, 567,000,000 of tons (2,000 pounds.) I have made no allowance in this computation for loss in pillars or waste in mining.

The western edge of the basin is disturbed by three narrow faults or fractures of the measures; they do not make much showing on the surface, but they cause the measures in their vicinity to be irregular, and will not be considered worth working while there is such a vast area of almost level or flat measures in the basin proper, to the east of them, and containing the same seams.

The Gould seam outcrops in these disturbed measures, but the lowest workable seam outcropping in the regular or flat portion of the basin, is the Wadsworth, which shows two feet nine inches at the surface outcrop and will probably prove to be three feet of good coal at some distance underground; the next working seam above this is the Beechtree seam. This seam, a few miles to the east near Dailey Creek, is three feet in thickness and of good quality. The next workable seam above this is the Coke seam, this one also near Dailey Creek, is three feet in thickness of good coal, with a good roof and has excellent coking qualities. The next workable seam above this is the Woodstock or Gholson seam; in this basin it averages from three to three and a half feet of solid coal of good quality for coke making, and locomotive or domestic purposes; it has a good roof, and around Blocton there is a large area of it nearly level. The next workable seam in this basin above the Woodstock is the Underwood or Thompson seam; this seam contains a solid bench of five and a half feet of good quality well suited to coking, steam, or domestic purposes. The following is a section of it:

[*Thompson seam*, in section 21, township 22, S., range 5, W.]:

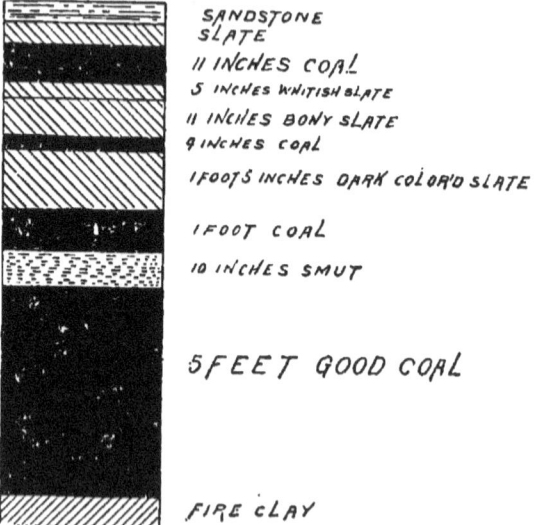

The Helena seam is the next workable seam above the

Thompson; it shows only two feet in thickness at the outcrop on the hill above the No. 2 slope in this basin. It may be larger in other parts of the basin, though the evidence elsewhere testifies to its gradually reducing in size towards the southwest end of the Cahaba Coal Field. In places through the field it is liable to be divided up into two or three benches, with slate intervening; in the Eureka basin it is solid.

For relative position of the seams in this basin, see the *Blocton Vertical Section*, the *General Vertical Section*, and the *Blocton and Montevallo Horizontal Section*, on the accompanying map:

The following two analyses of the coal of the Woodstock seam, were made by Porter & Going, Cincinnati, Ohio:

	Sample No. 1.	Sample No. 4.
Moisture	1.45	1.40
Volatile	32.21	34.05
Fixed carbon	61.83	60.30
Sulphur	1.10	1 14
Ash	3 41	3.11
	100.00	100 00

The following two analyses of the coal of the Underwood seam were also made by Porter & Going, Cincinnati, Ohio:

	Sample No. 2.	Sample No. 3.
Moisture	1.70	1.50
Volatile	32 21	30.95
Fixed carbon	60.02	61.72
Sulphur	.82	1.13
Ash	5.25	4.70
	100.00	100.00

The following analysis of the coke from the Underwood coal was made by Alfred Gaither, Chemist, Philadelphia, Pa.:

Volatile	4.508
Fixed carbon	87.607
Sulphur	.745
Ash	7.140
	100.000

The following analysis of the coke from the Woodstock

and Underwood coals mixed, was made by chemist of the Talladega Iron and Steel Company:

Moisture	700
Volatile	.925
Fixed Carbon	88 358
Sulphur	1.217
Ash	8 800
	100 000

The following analysis of the coke from the Woodstock and Underwood coals mixed, was made by John Fulton, General Manager of the Cambrian Iron Company, Johnstown, Pa., from samples taken from twenty-four ovens:

Moisture	08
Volatile	1.11
Fixed carbon	90.48
Sulphur	.83
Ash	7 50
	100 00

The disturbed measures next to the northwest edge of the Blocton basin have a varying rate of dip of from six degrees to sixty degrees. The main part of the basin is nearly flat, the rate of dip varying from one degree up to fifteen degrees. The synclinal of this basin is wide and flat, and extends from the northeast end to the southwest end.

Around the Cahaba Coal Mining Company's mines the synclinal becomes divided by an anticlinal that shows itself between No. 1 and No. 2 mines, into two synclinals, extending for several miles in both directions. These synclinals are wide and almost flat, and embrace a large territory of nearly level measures. The inclination or fall of the synclinal line, of this basin, is from the northeast end to the southwest end. The base of the Millstone Grit, measured from a given datum line, has a lower altitude at the south end of both the Cahaba and Warrior Coal Fields, and a higher altitude at the north end of both Coal Fields than at any other point; consequently the large Montevallo Conglomerate, the cap rock of our Alabama Coal Measures, is visible at the surface at the south end of both coal fields, which can be seen at the shoals in the Warrior River between Tuscaloosa and Northport, and in the Montevallo

CAHABA COAL FIELD : BLOCTON BASIN. 117

Basin over the "Aldrich Slope," The base of the Millstone Grit showing itself in the tops of the mountains where the measures have a very light dip, at the north end of both coal fields, more especially the Wearior.

Prior to 1884, there had been no mining done in this basin; in that year, the Cahaba Coal Mining Company first began to sink their Slopes and construct their nine miles of railroad from Woodstock, on the Alabama Great Southern, to their mines ; though they have now in this basin nearly twenty miles of railroad of main line, branches, and sidings ; they have increased their mine openings until they now have ten mines opened up in this basin, mostly slopes, the others are vertical shafts and drifts , their output has increased at about the same speed as the Pratt Mines, did in the same space of time after first commencing.

This company have some 450 coke ovens of the bee-hive pattern, well constructed, and with the latest improvements. They are intended to supply the furnaces at Anniston with coke. The coke is of excellent quality.

CHAPTER XIV.

ON MINING.

In our methods of mining the coal seams of Alabama, where the rate of dip is less than ten degrees, we have adopted for the past thirty or forty years, the cars and system very generally used along the Monongahela River, Pittsburgh, Pennsylvania, and for seams having a rate of dip of from twenty-five to sixty degrees, we have adopted the methods generally used at the Anthracite Mines in Pennsylvania. For distinction we will name the first one the "Monongahela Method," and the other the "Anthracite Method," and for the rates of dip above mentioned, they are the best methods known, but they do not work well in seams having a rate of dip between ten and twenty-five degrees.

In seams having a rate of dip from forty to sixty degrees, it has been our custom to drive the rooms square off from the gangway, up the "rise" of the seam, and have the coal to run down the shute into the tram at the bottom of it; with this rate of dip the shute does not require planking at the side or bottom to make the coal run, and by keeping the shute full, except three or four feet working room at the "breast of the room," there is very little coal lost by pulverizing in its descent down the shute, as by that method it descends by slow settling in proportion as it is allowed to run into the trams at the bottom; this method miners designate as "working it on the run."

In seams of from thirty to forty degrees rate of dip, the miners are compelled to plank the sides of the shute to some extent, in order to enable the coal to slide down without assistance. In seams of from twenty-five to thirty degrees the coal will not descend in the shute unless the sides of the shute are partly planked, and the bottom covered with sheet iron. In working our seams, having a rate of dip of ten degrees or under, with the Monongahela ton car we are compelled to drive our rooms diagonally to the di-

MINING.   119

rection of the gangway, unless the rate of dip is less than four degrees, in that case the rooms may be driven "square up the pitch." For seams of from sixty to twenty-five degrees and from ten degrees to flat or level, the Anthracite and Monongahela methods suit very well, but for seams having a rate of dip of from ten to twenty-five degrees, they entail an additional expense in getting the coal to the gangway ready for hoisting; for convenience we shall name this rate of dip the "medium dip."

It has been hitherto our practice to adopt the "Monongahela Method" with ton trams, where the rate of dip is from ten to seventeen degrees, driving the rooms diagonally from the gangway, and have the miners bring their loaded cars down to the gangway, go back empty handed and have the trammer to take the empty cars up to the room breast by mule power; or else have the miner to go through the heavy strain of pushing the empty car up by hand. The mule power method, though necessitating two journeys along the room road, to accomplish the output of one car of coal, is the most satisfactory to the miner and most economical to the mine proprietor; in making a fair count of the cost of each method, the man power is certain to cost the most. In mining thin seams, small light cars are often used, that can be pushed up the room by man power with less strain to the miner than when using the one ton car. I have often used this method myself, and in all probability the Montevallo Coal & Transportation Co., are now using it, still it is glaringly evident, that man power applied to its utmost strength, is the costliest method of moving coal from the "room breast" to daylight.

In mining seams of from seventeen to twenty-five degrees rate of dip, it has generally been our practice to adopt the "Anthracite Method," and either plank the bottom and lower part of the sides of the "shute," or plank and sheet iron the bottom. In this case, even with these aids, the coal will not run of its own accord, consequently it requires to be pushed down the length of the shute by the miner or the assistant trammer. When the room is worked up a considerable distance from the gangway, this becomes a costly method of moving the coal from the "room breast" to daylight. I have given considerable attention in the past thirty years, to the difficulties encountered in conveying the "medium dip" coal from the "room breast" to daylight; twice in this period I have tried to solve the problem, by devoting several months to the examination of the methods used in the "medium dip" seams of England, Scotland, and Wales. I also made a further effort on the Continent, but my knowledge of German and French was so limited, as to prevent my discussing the matter satisfactorily with the managers in charge of the works. As the result of these efforts I have been brought to suggest and recommend some (at least to me), new methods, though not an entire "cut and dried" solution of this problem, ready to apply to our Cahaba seams.

The trams or mine cars used in Europe are, in nearly every case, smaller than ours; the reason for making them so, in most cases, is an effort to reduce the enormous first cost of their deep shafts, by having a small shaft area, thus leaving but a small space for their mine cars or cages and pumpway; their small mine cars also suit the large number of boys they have employed in their mines. It would be bad policy for us to adopt their small cars in the Cahaba Field, as we have no very deep pits to sink, and our percentage of boys employed is very much smaller than theirs, also our miners are accustomed to handling one ton cars, or cars having a capacity approaching a ton. I have also examined the methods of mining the "medium dip" in other places where opportunity offered, finally arriving at the conclusion that our best policy is to hold on to our one ton cars, and work the "medium dip" seams horizontally.

The most improved method of tramming and removing the "medium dip" coal, that has come under my observation, is that mostly used in the county of Lancashire, England. The diagram opposite is the ground plan showing endless wire rope haulage, and section of it, and I shall designate it as the "Lancashire Method."

It must be borne in mind, however, that in that county the system of "underground wire rope haulage" is in almost universal use. This "Lancashire method," is an application of the "endless wire rope haulage"; the slope is double

LANCASHIRE METHOD OF WORKING SEAMS WITH RATE OF DIP OF FROM 10 TO 25 (DEGREES) HORIZONTALLY BY MEANS OF THE ENDLESS WIRE ROPE SYSTEM OF HAULAGE.

GROUND PLAN SHOWING ENDLESS WIRE ROPE HAULAGE.

tracked, the endless rope ascending up the middle of one track and going down the middle of the other. The room roads connecting with the slope on each side, are opposite each other; and in both tracks there are level spaces opposite the room entrances, to facilitate the pushing the mine car under the rope towards or from either track.

The method of hitching the mine car to the wire rope is by means of two chains (one at each end of the car) resembling our trace chains, only with shorter links is the hitch to the rope is made in the same time (about one second), that the other end of the chain is hooked to the end of the mine car. In hitching to the wire rope they give the end of the chain a sharp swing around the rope, and after the hook has made two rounds, they catch the hook with the other hand and put it over the chain. When the slope is made down the "dip," then full cars are hitched to the ascending rope, but when the slope is made up the rise of the coal, then the full cars are hitched to the descending rope. The system is used for lowering loaded cars to a lower gangway, and for hoisting them to a higher gangway, and it works well at either, and by this method in circumstances that suit it, coal can be conveyed a given distance underground at less cost than by any other appliance. The Lancashire method just suits their mine cars; their endless ropes have a continuous steady motion of 1 1-4 to 2 1-2 miles an hour without stopping the whole day; every miner is trained and able to push his car under the rope, and have it under way, without interfering with the car following after it.

Our cars are so much heavier than theirs that it would probably be impossible for one man to push them under the rope and hitch them quickly enough to keep them out of the way of the following cars. I am uncertain about the possibility of using the above described method with one of our one ton cars, so shall leave it to time, or some of our enterprising mine operators to decide its feasibility with the mine cars are now in use here.

Another system of mining the "medium dip" seams, or, more correctly, a combination of different and various methods now in successful operation in many old established mining districts, is, in my judgment, superior to any other method for seams having a dip of from 10 to 25 degrees from the horizontal : it is the best suited to our seams, our mine cars, to our miners, and to our "pillar and room" habits of working ; and, as it is a combination of methods partly used in one, and partly in other districts, we shall designate it as the " Combination method."

In this method, the system of conducting the underground workings is, to have but one single track slope driven in the direction of the dip. This we will name the drainage slope; the pumps being in a narrow air-way at one side of it. By this slope all the coal within its jurisdiction or territory will be drained, and it will also be the medium through which the coal and slate of the adjoining gangway end hoisting slope must be brought to the surface. All work in this system will be driven either horizontally or directly up the rise of the coal, (excepting the draining slope.)

The diagram opposite gives an outline of this system of working.

A pump and air-way is driven up at one side of the drainage slope, and hoisting slopes are driven up at suitable distances on each side of the drainage slope. In order to avoid the expense connected with long underground haulage, the rooms are all driven horizontally or nearly so, the grade of the room tracks must be laid to the proper inclination, by means of a tapering grade stick, with glass level imbedded in plaster of Paris, and adjusted to a three-eights grade (or 3-8ths of an inch to the hundred inches), or to such grade as the size and style of wheel used in mine cars may require.

The drainage slope will require coal pillars large enough for its permanent security. With this method a room can be advanced 150 *yards* with no more outlay of strength and muscle to deliver the coal and secure an empty car, than will be required to advance a room 150 *feet* diagonally up our "medium dip" seams, or in other words the miner can push his full car out, and return with the empty 150 *yards*, at less cost and exertion, than would be expended in the same work through 150 *feet* in the diagonally driven room up the pitch of our medium dip seams.

PROPOSED METHOD OF WORKING THE ROOMS HORIZONTALLY OF THE CAHABA COAL FIELD SEAMS, THAT HAVE A RATE OF DIP OF FROM 10 DEGREES TO 25 DEGREES

APPROXIMATE SCALE 500 FEET = 1 INCH

The grade stick can be so adjusted that the same muscular strength will be rexuired to push the full car down, as to push the empty car up, the only trouble being to put the grade stick on the track when laying it, and support or lower the ties until the bubble sets right. On the diagram the distances between the hoisting slopes are spaced in pannels of 900 feet, but that distance can be lengthened or shortened to suit the locality and the seam. The method of working "medium dip" seams, has less amount of narrow work to a given acreage of coal than any other method yet made known, excepting the "long wall" method, and before we can adopt the latter, we must reduce the size of our cars, and train and discipline our miners to work under a "sagging" roof, and if the "long wall" is the "withdrawing" kind, we must lay tracks along the "face or breast." The hoisting power at the top of each hoisting slope, can be either steam or electric motor connected with a central dynamo. If steam is used, the water would probably have to be piped from the drainage slope.

The long underground haulage is one of the chief drawbacks to our "medium dip" seam mining; in some districts the usual way to curtail that expense is to establish the underground wire rope haulage system. In the "combination method" the car bodies are strongly made wooden boxes of rectangular shape, of one ton capacity. These are detachable from the trams or trucks. In the rooms, the trams consist of a flat platform resting on the trucks, and of a size sufficient to hold a single car body. In the hoisting slope, the "hoisting" or "slope" tram consists of a long iron frame work on trucks, on which are constructed four steps or scaffolds, so arranged with reference to the slope of the track, as to have the floors of these platforms level at the steepest part of the slope. On each of these platforms is placed one of the detachable car bodies above referred to. The Diagram following p. 124 shows the construction of the "slope tram" with a car body resting on each of the four platforms, with ground plan of slope and room roads; also longitudinal cross sections of slope, showing hoisting tram.

The miner takes the empty mine car body from the "slope tram" and replaces it with a loaded or full one, signals to the engineman to hoist away, and goes back to his room with the empty car to get another car load, thus requiring only one trip to deliver a one ton car of coal. The "empties" are taken from the "slope tram," and the full car bodies placed upon the same, by means of an iron post crane placed at the angle of the room road. At this point the slope pillar, instead of coming out to an angle, is cut away sufficiently to give space called a "siding" (but which has no side track), for the empty car to be swung from the slope tram and held suspended out of the way, while the full car body is being placed upon the slope tram, after which the empty is swung still further around and lowered upon the platform of the room truck, from which the loaded car has just been removed. This necessitates at each room entrance, two cranes (upon a swivel post). The crane for the empty car body being of a lighter construction and placed on the upper side of the post; that for the loaded car, heavier and on the lower, or room entrance side. [See Diagrams, one opposite, and two following p. 126.] From the end of each of the cranes there is suspended by a swivel joint in the centre, a light beam of the length of a car body. This beam has a small pulley at each end, over which passes a wire cord terminating in a hook and fastened at the other end to running nuts on a double screw, actuated by a crank, on the same principle as the screws of the log carriage of a circular saw. This arrangement is for raising and lowering the car bodies. The screw for the "empties" is coarser, giving a a more rapid lift, than that for the loaded cars. The second diagram opposite p. 126, shows the arrangement of the screw threads and crank for raising the mine car body from the room tram or from the slope tram. A catch lever is arranged at the side of the slope track opposite the room road, to enable the miner to stop the "slope tram" at either one of the four platforms or scaffolds on which the mine cars rest in their transit up or down the slope. The mine car body is raised up from the tram truck, or up from the "slope tram" by means of a screw, which the miner turns by a crank as above described; the screw, when rotated, pulling a light wire cord above described sufficient to raise the car body a few inches, by a few turns of the crank; the crane is then

swung around and the mine car body let down by means of the same screw either on to the room tram or on the "slope tram."

In this method the engineman alone takes the place of all the trammers who, in other methods, are employed in bringing coal from the rooms or "breasts." In this method of mining the "medium dip" coal, there is a less amount of narrow work in the form of gangways and air courses, than in any of the usual methods; there is a much lighter force of trammers needed, and especially there no coal rakers, killing time in the shutes, in their dallying efforts to get the coal down the shutes to the gangway.

If the Pit Head Frame and loading shute and screws are properly arranged, the "medium dip" coal can be mined by this method at a very little if at all higher cost than the coal of the flat seams. The preceding diagram shows a section along the hoisting slope, giving an outline of the "slope tram," with form of the platforms or scaffolds for holding the mine car bodies; also a ground plan of the hoisting slope with its connecting room roads, and sidings for empty mine cars; also the position of the iron post cranes for receiving and delivering the mine cars.

The first duty of the miner on arriving at the slope from his room with his full mine car, is to signal to the engineman by means of the annunciator, that his number requires the "tram slope" with empty car, and is ready to deliver a full car; the engineman's duty, after acknowledging receipt of this order, is to signal back to the number at which he intends to stop his "slope tram," that he is going to stop at that point. As the slope tram nears this place the engineman causes it to move slowly in order to give the miner the opportunity of seeing which platform of the tram holds an empty car body, and of stopping it, by throwing up his catch lever, so as to bring this platform and empty car body exactly opposite his room track. He then removes the empty, and puts on the full car body and signals to the engineman to hoist away. He then swings the empty car around upon the room tram, pushes it back to the breast to be loaded again. The signals between the miner and the engineman must be the "electric," each miner having a wire to himself, with an electric light at each crane during working hours. The ends of the room tracks must be curved up so as to prevent the mine car from ever running into the slope. The first diagram opposite shows a section along the slope, and across the room entrances and the entrance to two rooms.

In this method the amount of work necessary to fit up the post crane and make the siding, is not half that required to open a room and put in a switch in the ordinary level gangway. To fit up the post crane, all that is required is to dig a hole in the roof sufficiently deep to hold the head of the post and prevent its slipping, then dig another hole exactly under that one (by a plumb line), in the bottom slate, put in place the post which is of iron and in two parts fitting one into the other by a screw, and turn it after the manner of a jackscrew, until it presses sufficiently against roof and floor to prevent its moving. The remainder of the work consists in digging off the corner of the slope pillar sufficiently to make room for the empty car to stay out of the way of both slope track and room track.

The scaffold in the siding is not absolutely necessary, but a light one there would enable the miner to have an extra empty car. The end of the mine car body must have two hitching or hooking places, one at its top edge of the car for the miners use, the other about the middle of the end of the car for the top or bankman to hook to for dumping the coal on the screen. This method has the great advantage of allowing the mine car wheels to be fastened to the axle in both slope tram and mine cars. There are no curves to go around, therefore no slip of wheels; mine cars with wheels fastened to the axle, the axle itself rotating, will last probably twice as long as those that are loose and have the axle bolted to the bottom of the cars; they also run much lighter and keep the proper gauge much longer. The second diagram opposite, giving a section along the room roads and across the hoisting slope, shows the arrangement of the screw threads and crank for raising the mine car body from the tram truck or from slope tram.

In this method the Engineman must have in front of him (with the end towards him, and its lower edge about eight

METHOD OF WORKING STEEP DIPPING SEAMS HORIZONTALLY HAVING RATE OF DIP OF 10 TO 25 DEGREES.
SECTION ALONG THE SLOPE AND ACROSS THE ROOM ENTRANCES.

JULIUS BIEN & CO. NEW YORK

METHOD OF WORKING STEEP DIPPING SEAMS HORIZONTALLY HAVING RATE OF DIP OF 10 TO 25 DEGREES.
SECTION ALONG THE ROOM ROADS AND ACROSS THE HOISTING SLOPE

feet above the floor), a cast drum with large thread or spiral cast on it, with the numbers of the different rooms in large figures painted on the spiral, so that the pointer will show him the exact place to a few inches, where his "slope tram" is, in order that he may run slowly when approaching the entrance to a room where he is to stop for a loaded car to be added ; this cast drum must have a geared connection with the hoisting-drum shaft. The Engineman must also have the number of the rooms close to his hands, so that he can arrange them in the order in which the calls from below are made, and remove them as the orders are filled. The collection of wires extending from the Engineman to the entrance of each room, must be bundled or twisted together and wrapped with thin sheet lead or tarred cloth, to prevent corrosion from exposure to dampness. In this method the wire rope has no sharp corners or small pulleys to drag around, and will consequently escape the breaking and tearing of wire strands so common where the ordinary hoisting rope drags the mine cars out of the gangways.

To facilitate the quick delivery of the loaded cars at the top of the slope, the upper part of the slope track, (that portion next the large rope sheave), must be double tracked and be movable, so that the full cars when they arrive there, can be pushed to one side, the same motion bringing the track with slope tram containing the empties in line with the slope, so that the engineman is not delayed, but can let down the empty cars while the top men are emptying the full ones. Three tracks of wide guage are requisite for the screening and loading shute, one for "lump" or "run of the mine," one for "nut and slack," and one for slate.

If this method, with the necessary machinery, were in common use, it is probable that it would be used for "medium dips," of even from five to thirty degrees. In cases where the dip of the seam is irregular, and becomes too flat to allow the "slope tram" to descend and overcome the drag of the rope, a light tail rope would have to be used. In this method the "long wall system" could be used to some extent, but considering that we use a ton tram mostly, and a kind of room track, different from that usually employed in long wall mining, and that our miners are mostly accustomed to the "pillar and room" system, it is probably best to adopt it only where the roof is good, the floor not too soft and apt to "swell up," and where there is abundance of hard gob material to give some support to the roof. In Warwickshire, England, they mine their medium dip coal (from fifteen to twenty degrees), by the "long wall drawing method; a full description of which is given by William S Gresley in the Engineering and Mining Journal of August 17th, 1889, and I have no doubt but that it is the most improved method of mining the medium dip seams now in use in Warwickshire, and that it suits their condition of mining matters, is very evident. In the first place, they have to go to their boundary to commence the withdrawal of the coal, while in our case, most of our mine proprietors know that even their grandchildren will never extend their underground workings to their boundaries; in the second place, their small square sided mine cars can be taken between the props and the face of the coal, much more readily than our cars of the Monongahela pattern; in the third place their room tracks have a sawed flat tie, of one and a quarter inch thickness, with the ends of the rails locking into one another, and with holes in the ties that keep their rails in guage, so that they can move their tracks along the breast, while we are knocking out the wedges, or drawing the spikes of ours.

While in Europe, some ten years ago, the underground system of wire rope haulage received my attention, and I devoted several months to a thorough examination of the various methods of using it, and found its greatest development in the Wigan district of Lancashire, England. It was no new experiment to them, as several of the mine superintendents informed me that they had abandoned the use of pony or mule and trammer, twenty years previous to the time of my examination, or now a generation ago. The proprietors and managers showed me ropes that they were using then, that they had been using constantly underground the thirteen years prior to that, the rope still good.

Their underground haulage ropes are made of steel wire, with a hemp core. In one pit that had a regular output of

800 tons per day, they had but one mule or pony in the pit at any time, and it was in charge solely of the repair man, to haul about their props and repair material. The mine proprietors informed me that, if they were to fall back to the old style of pony (or mule) and trammer to haul their coal to the pit bottom, that it would ruin their business, for they could not compete in that case with their neighbors using the underground wire rope haulage.

They had passed the experimental stage long ago, knew at a glance the kind of pulley or sheave, in their great variety, that was essential to enable the rope to work well in the thousand and one difficult localities of their gangways and slopes. They have a large number of variously shaped pulleys, and modify their methods of using their wire ropes to suit the varying circumstances that surround them. They have two methods of conveying the power down their pits to their systems of wire rope haulage : the one in most general use is compressed air from air compressors at the top, to compressed air engines near the bottom of the pit; the other method is to have a steam engine at the pit top, geared as to give a slow motion to a large broad grooved sheave, having two or three wraps of the rope around it. This is carried down the pit to the pit bottom, and from thence to the various parts of the pit, where the power is needed ; this rope is driven at a speed of 1 1-4 to 1 1-2 miles per hour ; this manner of conveying the power suits the endless rope the best, while the compressed air engine suits the "tail rope method," or any style where quick motion is required. The leading systems in use mostly are :

    The Endless Rope System.
    The Tail Rope System.
    The Simple Engine Plane or Slope.
    The Gravity or Self–Acting Plane or Slope (called when the rope is endless, an "endless jig.")

These systems are all modified to some extent to suit the varying circumstances. The endless variety of their appliances to prevent their ropes from rubbing, convinced me that they paid close attention to wear and tear of their underground ropes, some of which are over two miles in length. Their endless ropes run slowly; 1 1-4 miles per hour is deemed best, 2 1-2 miles per hour being their highest speed. Their common hoisting speed in pits of a quarter of a mile vertical depth, is one minute for the quarter of a mile; this includes the slow run near top and bottom; one of their tail rope trains of about ten cars, passed me in one of their gangways at a speed of ten miles per hour; this rather surprised me, but I was more astonished on noticing that the boy in charge of the train was stretched out at full length on top of the last mine car, his head and back not over a foot from the roof; his only chance to stop the train was to jerk the signal wire at the side of the gangway, the engine being a half mile away.

I was informed by the mine managers that ten miles per hour was the ordinary speed of their "tail rope trains" in the middle of the haul. Yet with all the advantages and economy of the system of underground wire rope haulage, the lack of machinery and appliances, the absence of labor skilled and trained to handle and use it, will no doubt cause our mine managers to hesitate considerably before adopting it, but should any of them decide to adopt it, their best plan would be to go and see it in operation, examine the different systems, and study the various changes made in the use of the appliances to suit the different conditions and circumstances, then make arrangements to secure the machinery and appliances as needed, in the section of country where wire rope haulage is well understood and extensively used; then begin with the simplest and easiest form of wire rope haulage and increase gradually as the laborers become more skilled and trained. To begin to adopt it in its more complicated forms, perhaps might result in failure and disaster. For conveying power to the "tail rope system," or any other quick motion system of underground haulage, where it is a long distance from daylight, the dynamo, electric wire and electric motor is superior to compressed air or any other method, and more economical. The electric wire will yet supply with power all mining pumps and wire rope haulage systems, that are situated a long distance from daylight.

Instead of copper wire, iron rods of 5 or 6 times the sectional area of the copper wire, will answer equally as well or better, for conveying power underground.

For conveyance of power from the surface to endless rope systems that are not very distant from daylight, the rope itself, driven by a steam engine at the surface, and moving at the rate of $1\frac{1}{4}$ miles per hour, is the cheapest and most economical conveyance of power to underground haulage. There is nothing more certain than that in the future, wire rope haulage power and the electric power, will be used extensively in underground mining operations. It may be safer to be wary and move cautiously in their adoption, increasing their use gradually, still it is only a matter of time as to their general adoption.

In the gangway of our Cahaba Field mines, the overhead electric wire would be too dangerous if not insulated. In fact all electric wires of high voltage placed in mines should either be insulated, or placed in narrow channels so that there would be no possible chance of the minor coming in contact with them.

The storage battery with electric motor (thus doing without wires), is the best and safest method of underground electric haulage in gangways that are level, or nearly so.

In conveying power to pumps or drills, there is no necessity for using any but insulated wires.

PART II.

GEOLOGICAL STRUCTURE AND DESCRIPTION OF THE VALLEY REGIONS ADJACENT TO THE CAHABA COAL FIELD.

—BY—

EUGENE A. SMITH.

CONTENTS.

I.—Origin of the Rocks of the Cahaba Coal Field and adjacent Regions, and the Agencies which have brought them into their present position..............................Page 137

II.—Classification of these Rocks, and their distinguishing Characters ....Page 146

III.—Distribution of the Rocks of the different Geological Formations in the Valleys bordering the Cahaba Coal Field.Page 159

I. ORIGIN OF THE ROCKS OF THE CAHABA COAL FIELD AND ADJACENT REGIONS, AND THE AGENCIES WHICH HAVE BROUGHT THEM INTO THEIR PRESENT POSITIONS.

The map and sections of Mr. Squire exhibit the structure of the Cahaba Field in sufficient detail, but a few words explanatory of the relations of this field to the others, and to the valleys lying between them seem to be required.

It is the commonly received opinion among geologists, and an opinion capable of demonstration, that the older stratified or bedded rocks of the Appalachian region of the United States, in which is included Cahaba Coal Field and the regions above alluded to, were formed partly out of the detritus of a previously existing land mass lying to the eastward of the present shore line of the Atlantic ocean, and partly out of the calcareous and siliceous matters accumulated through the agency of living organisms, in the depths of an inland sea which formerly occupied the position of the greater part of the present United States. This detritus, washed down by rains and transported by rivers, was finally spread upon the floor of this inland sea. Naturally by far greater part of this land waste would be deposited close to the shore line, while only the finer sediments such as silt and mud would be held in suspension long enough to be carried far out to sea and be deposited there, and in the clear and moderately deep waters of the sea at a distance from the shore would flourish the corals, and other organisms that formed the limestones and part of the chert or siliceous matters. If the floor of this interior sea remained stationary while receiving these sediments, it is easy to see that it would very soon be silted up by the washings from the land, and that no great thickness of variety in the sediments would be seen at any one place; we should not find, for instance, alternations of limestone with sandstones and conglomerates, while, in point of fact, the sediments which make the rocks of these older formations are many thousand feet in thickness and consist of sandstones, conglomerates, shales, and limestones in many alternations.

All this is clear demonstration that the floor of the sea did not remain stationary during this period, but subsided, —at least to the extent of the thickness of the sediments accumulated upon it,—not steadily and continuously, but with many pauses of downward movement, alternating even with movements in the opposite direction, which went so far at times as to bring parts of the sea bottom above the water, and to afford the requisite conditions for the accumulation of those immense beds of vegetable matter that constitute the seams of coal.

In the manner above sketched, there were accumulated upon the floor of the interior sea, and in the marshes and peat bogs of the land, and in the estuaries of the rivers, during a period of whose duration we have no means of making a definite estimate, beds of gravel, sand, mud and limestone, and coal beds, of varying thickness according to position; from 40,000 feet near the margin of the sea where the greater part of the land waste was deposited, to 4,000 feet further out to sea where the materials deposited were mainly calcareous and siliceous. These beds contain the remains of the animals and plants that flourished upon the land or in the waters of the ocean during the period of their accumulation, and when consolidated and elevated above sea level they constitute the rocks of the various geological formations. These rocks and their contained organic remains, have been objects of study and investigation among geologists for many years, and as one of the results of these investigations, they have been classed together into a number of great groups having certain common characteristics of mineral composition and fossils. The names of these great geological groups or formations beginning at the lowest and proceeding upwards, are Cambrian, Silurian, Devonian, and Carboniferous. The maximum thickness of the rocks of these formations, as they are displayed in Alabama, may be approximately given as follows: Cambrian 10,000 feet;

Silurian 5,000 feet; Devonian 100 feet; Carboniferous 6,500 feet, making in all not less than 21,600 feet.

We must next endeavor to explain how these beds have been elevated above the sea so as to become a part of the dry land, and how they have been brought into the positions which they now occupy. As originally deposited, we may infer that they were spread out upon the floor of the interior sea in sheets or strata, which, allowing for the slopes and inequalities of the sea bottom, and the greater thickness of the deposits near the shore, were in approximately horizontal position, and if they were brought up above sea level by some gradual and uniform motion of elevation, we should have a condition of things such as prevails in the lower part of this State, in the territory made by the newer formations Cretaceous and Tertiary, viz., the beds thus elevated would be nearly horizontal, but with a slight slope or dip towards the sea, or towards the northwest; there would be no mountains or great inequalities of surface except such as might be produced by the erosion of rains and running waters, and at any one place only a very few feet in thickness of strata could thus be exposed. We also see to the northwest of the region with which we are here concerned, in Alabama, Tennessee, Kentucky, and beyond, approximately level or horizontal strata into which erosion has cut only a few hundred feet, and exposed only a few hundred feet of the uppermost beds. On the other hand, we notice running diagonally through the upper half of Alabama and thence northeastward through the other States to Canada, a belt of country perhaps to 150 to 200 miles in width, the strata of which are seldom in horizontal or even approximately horizontal position. They are inclined to the horizon at varying angles, being sometimes even perpendicular; their outcropping edges may be followed for many miles in a northeast direction; the lines of outcrop of the edges of different beds are approximately parallel with each other, and by crossing over these outcrops in a direction at right angles to their trend, i. e., from southeast to northwest, we may pass in succession over the strata of the whole series of geological formations from Cambrian up to Coal Measures, and all within the distance of a few miles. A further inspection of these rocks will show us that they have not only been tilted up but have been crushed together, and folded in a very complex way, and that rocks which are widely apart in the geological scale, are often found in direct contact. We shall see, moreover, that these disturbances are more profound along the southeastern part of this belt, and constantly diminish in intensity as we go northwestwards, so that the strata even in the northwestern part of this State, are thrown very little out of their originally horizontal position. It is evident therefore, that the strata of this region have been subjected to the action of some other force than one by which they were merely gradually elevated, and that whatever may have been the origin and nature of this force, it was much more pronounced in its effects along the southeastern border of the disturbed region, than further to the northwest.

The same pecularities of structure and attitude characterize the rocks of the whole Appalachian region from Alabama to New York and beyond, and these matters have been closely and carefully studied by many of the best geologists of the country, the brothers Rogers, Safford, Lesley, Dana, and others; most of the peculiarities of Appalachian structure have been described, and satisfactory explanations of the approximate causes of these peculiarities have been given.

No one who will carefully examine the positions of the various rocks exposed, for instance, in Jones' valley, can fail to see that these rocks have been pushed up, in such a way as to cause their broken or exposed edges to trend or run in the general direction of the course of the valley, i. e., northeast and southwest, and that most of these rock ledges show a dip or slope towards the southeast. This position of originally horizontal beds could be brought about only through the action of some force coming either from the southeast or from the northwest, and compressing them together in that direction into much narrower limits than they originally occupied, and this compression into narrower limits could take place only by the strata being thrown into a series of wrinkles or folds, or by their being rent apart and one side slipped up over or past the other. There are many reasons for the conclusion that the force in question came from the southeast rather than from the northwest, one of these reasons among many, as already said, being that the intensity of the disturbance constantly diminishes as we go from southeast to northwest.

The varying degree of deformation of the strata by varying amounts of compression can be imitated on a small scale and illustrated by pressing together sheets of cloth of clay or other plastic material.

If we place on a table a number of sheets of flexible cloth piled one upon the other like the sheets in a pad of paper, and fixing one edge of this pad, push or slide along the table the opposite edge towards the fixed edge, we shall see that a number of wrinkles will be at once formed across the sheets of cloth at right angles to the direction of the compression. If we continue to press the edges of the sheets towards each other, the arches will rise higher and higher, and begin to lap over in one direction, which, in the majority of cases, will be the direction towards which the shoving force acts. In a few cases the troughs will be shoved under the arches and the folds will lap over in the opposite direction.

Now, if we study closely the folds or wrinkles into which the strata of the region about which we are now writing have been thrown, we may easily recognize the very same arrangement. There are simple folds or arches, with almost equal slope on each side of the crest line, but these are rare; there are folds in which the arches have been pushed over towards the northwest, making the slope on that side steeper than on the southeast, these are very common; there are folds which have been pressed together so that the two sides are about parallel, and then lapped over to the northwest, these are also very common. On the other hand we find folds in which the troughs have been shoved under the arches so as to cause the steeper slope to be on the southeast side, and when this movement has gone on far enough the arches have the appearance of having been lapped together and pushed over towards the southeast by a force acting from the northwest; these cases are by no means so common as the others, yet we see in Murphree's Valley and a few other places good illustrations in point. After the folds have been pressed together and lapped over to one side, no further yielding to the compressing force can take place except by the giving way of the strata and the sliding of one part over the other, in other words, by the breaking apart and piling up of the beds. Now when a break occurs in a fold of the usual type, i. e., one which has been pushed over to the northwest, it is along the crest of the arch where the strain has been greatest, and the southeastern side slips up over the northwestern. Faults of this kind are usually designated as *thrust faults*, and the displacement sometimes goes so far as to shove a great body of strata over other beds for many hundreds of feet, and in some countries for miles even. In folds of the other class named, i. e. where the troughs have been shoved under the arches, the break occurs near the bottom of the trough, and the strata on the southeast of the line of fault are slipped under those on the northwest. The general effect of this kind of slip or fault is the same as if the compressing force had come from the opposite direction and had produced a thrust fault of the ordinary kind. These are also thrust faults, but to distinguish them from the normal type of thrust faults they might perhaps be called *reversed thrust faults*. In Murphree's valley and west of McAshan mountain, we have fine illustrations of this type of structure. In all these thrust faults we have either the older beds slipped up over newer ones, or newer ones shoved under the older, in either case bringing about a reversal of the natural arrangement.

But there is another kind of reversal. We have seen that all our Alabama thrust faults are, in their origin, folds in which the strain of the compression has been carried beyond the limits of endurance of the strata, and hence when the break occurs along the crest of an arch of the typical sort, the gently sloping beds of the over-riding side will slip up over the steeply inclined or even overturned edges of the beds of the overridden side, the inclination of the edges of this side depending upon the degree of overpush or over-lap of the fold, and it may be quite possible that in the movement of the one series of beds over the other the edges of the underlying series may by friction be bent still further in the direction of the thrust. In this way the upturned edges of the overridden side may be carried beyond the perpendicular and be actually reversed. Instances of this kind are common enough; the cross section given on another page shows it, particularly on the southeastern border of the Cahaba field, and on that of the Warrior field.

In a similar way, when the break occurs near the bottom of a trough that has been shoved under an arch, the edges of the under-shoved set will be bent or turned back more or less, and this also may go so far as to cause a reversal. We see this along the eastern edge of Murphree's Valley almost its entire length.

So far as I know, all the Alabama thrust faults have highly inclined or overturned strata on one side of the faults, and these vertical or reversed beds will be on the northwest or southeast side of the fault according to the character of the fault, whether a typical or a reversed one. In the great majority of cases the vertical or overturned strata are on the northwest side, for the reason that the great majority of the faults are typical ones.

Usually the upturned edges occupy only a narrow belt, because part of them are generally below the surface, in the fault, and covered by the overriding measures; but we have one magnificent example of the reversal of a great series of beds, in the overturned measures of the lower part of the Cahaba field, west of Montevallo, for here is a strip of the Coal Measures, two miles wide and six or seven miles long, pushed over beyond the perpendicular to an angle of 60°, and at the border of this strip we have the instance of the complete overturning of the measures and the gliding of the Cambrian strata over them, described in detail in another place and illustrated by a photographic view.

The folds above spoken of are not symmetrical waves with crest and trough of equal width, but, as may be seen by any map of the Appalachian region, consist of rather narrow crests, with wide troughs between, in which the strata are either approximately horizontal or only slightly undulating. These troughs, or the most important ones, with raised edges and with the strata sloping from each side towards the central line (*synclinal*), are the coal fields, which have to greater or less extent resisted the denudation which carried away so much material from the intervening crests. It may be asked why the strata along the crests of the folds were so much more completely removed than from the troughs. One reason of this may be found in the fact that the strata along the crests would be more or less torn and disrupted from the strain of the folding, while those of the troughs would be more or less compacted by compression. This, along with other causes, has led to the formation of anticlinal valleys, that is, of valleys which have been eroded out of the tops or crests of anticlinal folds, and of this character, more or less masked by faults, overlaps, and other complications, are the valleys above named which border the Cahaba field. In all these valleys, the strata were raised up first into ridges with perhaps originally somewhat equal slope both ways, northwest and southeast from the central line (*anticlinal*); with increase of pressure the folds were pushed over towards the northwest; compressed together and lapped over to the northwest; broken apart and slipped; and finally by erosion, worn down into valleys in which now only the projecting edges of the strata are seen. These, by their relative position, give us the clew to the structure. When the strata were thrown into waves by the compressing force above spoken of, the crests of these waves were raised much above the level of the intervening troughs, and when, by subsequent denudation these arches were worn down to the general level or nearly to it, the lower strata of the arches were uncovered and exposed to view, usually in the form of projecting ledges in the case of the harder rocks, and of trenches in the case of the softer and more easily eroded ones.

In this way the strata of the different geological formations down to the lowest, have come to occupy the surface in these valleys, usually in strips or belts which run approximately parallel to the length of the valley, and which, in consequence of the anticlinal structure are normally duplicated, though as a result of faults they sometimes appear only once in a section across the valley, and sometimes where, as in Jones' Valley, the structure is a double anticlinal combined with faults, they are repeated a third time.

Illustrations of all three of these cases will be given in the special description of the valleys.

It seems hardly necessary to state in so many words that the strata of our different Coal Fields as well as of the geological formations that underlie them, were from their very mode of origin continuous, and that their present separation has come about through the foldings, faults, and denudations, which we have been describing.

We might infer that after the strata had been thus brought up and added to the land area, their subsequent history would be merely a record of gradual degradation and leveling down by erosion. But we have evidence in the lower part of the region shown on this map, that after this part of the State had been elevated and undergone the changes mentioned and attained almost its present configuration, it was in part again submerged below the water level, and was overspread by the washings from that part which remained above the water. Only in this way could the great beds of sand, clay, and pebbles which cover so much of the area in the lower portion of the map, have been deposited upon the ridges and the valleys of the old land surface. This submergence happened during the period termed by geologists the Cretaceous, which is comparatively modern as contrasted with the age of the formations above named. From the distribution of these beds we can see that the shore line during this time of partial submergence ran in a curve stretching from the northwestern part of the State to near the middle, at Columbus, Ga. To the west and south of that line the land sank below the water, while it remained above water to the east and north.

And still later, almost in modern times, geologically speaking, when the dry land area of Alabama had attained its present extent, and the surface had by long continued denudation acquired almost its present configuration, our State was again below water, receiving deposits of pebbles, sand and mud, which in the upper part of the State have since been in great measure been washed away again, but patches of which still remain often upon the summits of the highest hills. In the lower half of the State these deposits have been much less completely removed, but remain to form the great bulk of the soils of that section.

Of these later movements, it is not our intention to speak except in so far as may be necessary to explain the presence of these overlying surface beds which in places hide the formations with which we are now more particularly concerned.

II. CLASSIFICATION OF THESE ROCKS AND THEIR DISTINGUISHING CHARACTERS.

With this sketch of the manner in which the sediments were accumulated and afterwards brought up above sea level and into the positions in which they are now found, we may go on to speak of the distinguishing characters of the rocks with their contained fossils, of each of the great groups or formations Cambrian, Silurian, Devonian, and Carboniferous, and to note the minor subdivisions into which they may be conveniently arranged for purposes of study and description here in Alabama.

It would lead us too far to undertake to speak of the characteristic fossils of each of these formations, except to say that they are more unlike the forms of the present day, the further we go back in the geological scale, and the resemblance to living plants and animals becomes more and more pronounced as we approach the top of our geological column; but in all cases, in the formations with which we are concerned in the present report, the resemblance of the fossils to living forms is rather remote. This has led to the grouping of the four formations above named into one division which has been called *Paleozoic* (Ancient Life), in allusion to the want of resemblance to modern forms. Except at a very few horizons, fossils are not abundant in our Alabama Paleozoic rocks, and rarely come under the notice of the ordinary observer, yet to the student of geology they are of the very greatest value since by means of them it becomes comparatively easy to determine the relative ages of the different formations containing them, when the stratigraphical relations of these rocks are not readily made out. As an illustration of this I might say that there are many places in Alabama, and particularly in the region covered by this map, where the rock beds have been completely overturned, so that the older beds are on top of the younger. It would often be impossible to determine the relative ages of these rocks by their physical characters, and where they have been overturned their relative position would of course, be absolutely misleading if we judged by tne stratigraphical position alone; but as each of these great divisions has its characteristic fossils, these become in many cases our safest, and sometimes our only trustworthy guides in determining the age of the rocks in which they are imbedded.

Since all these rocks have been formed either out of the detritus or waste of previously existing land masses (conglomerates, sandstones, grits, shales and slates), or through the agency of living organisms, (limestones, flinty or cherty matters, and coal and all forms of bituminous matters), one would naturally think that it would be impossible to distinguish one sandstone or one limestone from another, or in other words to distinguish one of our geological formations from 'another by its lithological or rock characters. As a matter of fact, however, the field geologist, after a very few weeks or months of practice, learns to distinguish the different formations by their rocks, and hence the lithological characters are of almost equal value with the fossils in classifying our rock formations, and inasmuch as the fossils are nowhere very abundant, in the great majority of cases we make use of the lithological characters alone in studying and identifying the different geological formations.

It is easy to see that it is nearly impossible to describe the rocks of these older formations in terms which will enable the inexperienced observer to identify them, yet a short account of the prevailing characteristics of the rocks is necessary to the full understanding of the description of their distribution in the valleys. It must, however, be constantly borne in mind that the characters of the rocks of all these formations vary with the geographical locality, they being generally coarser in texture and more siliceous towards the east than further west. Thus in the Cambrian formation there are in the Coosa Valley beds of immense thickness of a coarse grained sandstone or conglomerate, which in the valleys further westward, such as Cahaba Valley and Jones' Valley, are wholly wanting. So also the shales of the same formation are sandier in composition in the Coosa Valley and more calcareous in the two other valleys named.

THE CAMBRIAN.—The rocks of this formation are conglomerates, sandstones and shales in the Coosa Valley region, and shales and shaly limestone in the valleys which occupy part of the area of this map. The maximum thickness may be put at 10,000 feet, but this great thickness is seen only in the eastern part of the Coosa Valley, while in Jones' Valley the thickness is probably less than half the above.

The sub-divisions of the Cambrian which we recognize in Alabama are, in ascending order, as follows: the Coosa Shales, the Choccolocco or Montevallo Shales, and, interbedded with the last named, the Weisner Quartzite.

Coosa Shales.—In the valleys here described the rocks are, commencing with the lowest, thin-bedded limestones with clay seams between; usually very greatly contorted and tilted at high angles. Where these rocks come to the surface there results from their decomposition a very stiff calcareous clay soil. These lands being very level and hence badly drained, are not much cultivated, and in Alabama are generally known as "Flatwoods." The town of Bessemer is upon one of these "Flatwoods" tracts, and similar areas may be seen between Bessemer and Birmingham, and northeast of Springville towards Gadsden, and in the immediate valley of the Coosa River up to and beyond the line between Alabama and Georgia. The shaly limestones that give rise to these "Flatwoods," we have called *Coosa Shales.*

Montevallo Shales.—Above these Coosa Shales we find a considerable thickness of sandy shales of a great variety of colors, such as olive, green, brown, chocolate, yellowish, etc. The original material was a calcareous shale, but at the outcrops the calcareous matter has mostly been pretty thoroughly leached out, and only the more siliceous parts left. These shales crumble up in places into small fragments about the size and shape of *shoe-pegs*. Sometimes they are more tough and hard, and, especially towards the east, assume gradually the characters of the semi-crystalline rocks, and it is capable of demonstration that some of the partly crystalline slates of the eastern part of the Coosa Valley are only the changed or metamorphosed representatives of this division, which has been called the *Montevallo* or *Choccolocco Shales* from the characteristic occurrences in those localities. In Jones' and Cahaba Valleys these do not play a very important part except in the lower part of the Cahaba Valley from Centerville up to Montevallo. Beyond this limit they outcrop only in narrow and comparatively unimportant belts. In the upper part of the Montevallo Shales we find beds of blue limestone and gray dolomite which are often difficult to distinguish from similar rocks occurring in the next overlying formation. In fact the line between the Shales and the Knox Dolomite is, so far as Alabama is concerned, rather an arbitrary one.

Weisner Quartzite.—In the Shales above described and most commonly in their lower parts, are found in the eastern part of the Coosa Valley great beds of quartzite and conglomerate many hundred feet in thickness, but often of very limited extent geographically. The quartzites always form high and rugged mountains sometimes stretching for miles in an unbroken range, but as often forming detached and isolated peaks, rising suddenly out of the plains and as suddenly sinking down to the same level. The "Mountain" near Columbiana, the Kahatchee Hills, Alpine Mountain, Mount Parnassus at Talladega, Cold Water Mountain and Blue Mountain near Anniston, Ladiga Mountain above Jacksonville, Weisner Mountain east of Jacksonville, are instances of occurrences of this quartzite. The Weisner Mountain above named has been best studied, and its stratigraphical relations to the Coosa Shales and to the Choccolocco Shales, most clearly made out, for which reason we have used the name *Weisner Quartzite* to designate this member of our Cambrian, which occurs interpolated in the Shales as local masses of lenticular shape and often of very great thickness.

Prof. Safford, of Tennessee, has given the name *Chilhowee* to similar great masses of sandstone and quartzite occurring in that State apparently below the Shales above named, which he designates as the Knox Shale and Sandstone. In Tennessee the distinction between the shale and the sandstone member of the Knox Group, can be consistently followed out, but it does not seem practicable in Alabama to separate the two, for beds of tolerably massive sandstone occur at many horizons, interbedded with the shales. So also, for the reason that in Alabama the great masses of quartzite do not occur at the base of the shales, nor apparently, at any definite horion in the same, we have not used Professor Safford's name Chilhowee to designate the rock. Similarly it appears necessary to adopt a distinct name for the thin-bedded limestones with clay seams, of our "Flatwoods," since they play a very subordinate part if they occur at all in Tennessee. As above intimated, the Weisner Quartzite makes no show in any of the region covered by this map, and it is mentioned here only to give completeness to our enumeration of the Cambrian rocks.

THE SILURIAN.—We have not yet in Alabama found it practicable to arrange our Silurian strata in more than three principal divisions, which, beginning at the lowest and coming upwards, are as follows: Knox Dolomite, Trenton or Pelham Limestone, and Red Mountain or Clinton.

Knox Dolomite.—This name has been given by Dr. Safford to a series of rocks occurring in the vicinity of Knoxville, Tennessee, and, inasmuch as the rocks of this horizon in Alabama are identical with those described by him, we have retained the name in the Alabama Survey. This is one of the most important and widely spread of our older geological formations and its characteristic rocks are magnesian limestones or dolomites, sometimes quite pure, but more often impregnated with siliceous matter. This siliceous matter is sometimes found as a sandy impurity in some of the dolomites, upon the weathering of which it becomes quite prominent. For this reason, many of the dolomite beds of the lower part of the Knox Dolomite, when exposed to the weather, show a rough sandy surface, marked by shallow cracks running in every direction as if the rock had been hacked with some cutting instrument. These purer and sandy dolomites, together with some beds of tolerably pure blue limestone, occur near the base of the Knox Dolomite, and are very closely related to similar beds of the Shale division already described. On the other hand, the siliceous matter in the upper part of the formation is usually found in masses of chert of concretionary origin impregnating the dolomite, and on the breaking down of these rocks under the action of the weather, the calcareous parts are leached out while the siliceous parts remain usually in the form of angular flinty gravel, which forms the very characteristic ridges of the Knox Dolomite. In the region covered by this map, we have found it convenient to distinguish the area underlaid by the lower and more calcareous part of the formation and that formed by the upper or more siliceous part. In the former, the weathering of the limestones and dolomites has given rise to the formation of gently undulating terranes with a deep red-colored sandy loam soil of more than average fertility, which is the base of the best farming lands in all these valleys. The red lands about Elyton, and in parts of Birmingham, and in the Alexandria Valley across the Coosa, are good examples. In the upper part of the Dolomite the cherty or siliceous matter is more abundant as a surface material than the calcareous, and the country is broken or ridgy, rather than undulating. Some of these flint ridges extend for long distances unbroken. Good examples are the ridges of the North and South Highlands about Birmingham. In fact this angular cherty gravel is found upon all the lands made by the Knox Dolomite, but is much more abundant and characteristic in the upper part. The Knox Dolomite contains very few fossils, and these belong to the Lower Silurian horizon of the paleontologists, but we have in the chert itself a characteristic by which we can as a rule distinguish it from the chert of other formations, that is, we find in most of it small angular cavities of clearly defined shape which are usually thought to mark the places once occupied by rhombohedral crystals of dolomite, subsequently dissolved out. Prof. Safford was the first to call attention to this mark, which we have found to be an extremely useful one. The Knox Dolomite as well as the upper part of the underlying formation seems to have held originally much ferruginous as well as siliceous matter, and we find throughout the region formed both by the Dolomite and the upper part of the Shale, beds of the brown iron ore or limonite, which plays so important a part in the economic history of all this region. The iron ore seems to have been derived from these older rocks. As instances of the occurrence of limonite banks connected with the Dolomite and Shale, I may mention the Edwards Ore Bank near Woodstock, the mines at Greely and Goethite, in Jones' Valley, and the great beds at Shelby over the Coosa. The great bulk of the brown ores of Alabama is from this horizon.

At the top of the Knox Dolomite, and belonging perhaps to the next succeeding division, there is a rather peculiar rock occurring at intervals along Jones' Valley and elsewhere. It is a *breccia* made up of angular fragments, chiefly of the chert of the Knox Dolomite, cemented together into a rock which is a good many feet in thickness. This rock, being made of fragments of the Knox Dolomite, is of course younger, though on account of its materials we have usually classed it along with the Knox Dolomite. It is seen in greatest volume in the Salem Hills southwest of Bessemer, but occurs upon the Flint ridge forming the North Highlands at many pionts, e. g. Birmingham and Gate City, and also west of Springville. It has been called the *Birmingham breccia* by Mr. Russell of the United States Survey, and *Salem breccia* by us in the State Survey. It is of interest as showing that a period of disturbance intervened between the time of the formation of the Knox Dolomite and that of the Trenton Limestone. We have not attempted to show on the map the occurrences of this rock.

Trenton or Pelham Limestone.—As its name implies, this division is mostly calcareous. It may be perhaps as a maximum, 800 feet or more in thickness, and varies considerably in quality, the lower part being ususally impure and shaly, while the upper part is mostly a pure limestone, often used for the purpose of making lime and as a flux in the furnaces. The lower part commonly holds great number of shells of *Maclurea magna*, which is a characteristic fossil of the Chazy limestone of the New York Geologists. The purer limestone above, is also quite full of fossils, which, as a group, are those of the Trenton limestone of New York.

In places, particularly in the region south of the Cahaba Field in Bibb county, the uppermost beds of this formation, above the purer limestone mentioned, are calcareous shales and shaly limestones, often full of the fossil forms known as *graptolites*. Where these thin-bedded shaly limestones occur abundantly forming the surface, cedar glades are quite characteristic.

The valley between the Cahaba and the Coosa Coal Fields shows a wide belt of Trenton limestone, which is particularly pure and well developed near Pelham and Siluria in Shelby county, and southwards. Near Pratt's Ferry on the Cahaba, and stretching thence northeastward there is another great belt of it, containing some fine marbles, which have in a small degree been worked at Pratt's Ferry.

For the sake of completeness, I might add that the phase of the Silurian formation to which Prof. Safford in Tennessee has applied the name of *Nashville*, has its representative in Alabama though not within the area shown on this map.

The Clinton or Red Mountain Formation.—This is the third and uppermost of the divisions of the Silurian which we make in this State. The mass of the rocks of the Red Mountain are sandstones and shales, which show a great variety of color, yellow, red, brown, chocolate, and olive green, in this respect resembling the Montevallo Shales. Along with these are some calcareous and ferruginous rocks, the latter passing into beds of red iron ore, made up of small flattened nodules, shell casts, etc., of ferric oxide. In many places, where mining has penetrated the ore bed beyond the reach of atmospheric agencies, the ore is seen to be quite calcareous; in fact, a kind of highly ferruginous limestone, which, when used in the furnace, often contains lime enough to flux the ore. At the outcrop the ore is seldom calcareous, though often sandy. So far as I know there has been no very satisfactory explanation of the mode of formation of this ore. It is of very variable thickness up to twenty feet, and is in more than one bed. It is a remarkable fact that while near Oxmoor the ore is some twenty feet in thickness, just across the Cahaba Coal Field in the Cahaba Valley about six miles distant, the Red Mountain, or rather its representative, contains no ore at all in the greater part of its length, nor does it seem to contain any of the Clinton rocks. As is well known this formation furnishes the greater part of the material used in our furnaces. In places, the ferruginous limestone of this formation would make a fine building stone, and the same is true of the sandstones. It would be difficult to give the average thickness of the Red Mountain rocks proper, in the region of the present map; 100 feet might perhaps be a fair average, for the Red Mountain as a topographic feature, is made up of the rocks of different ages, Trenton, Clinton and Sub-Carboniferous, together with the usually very thin black shale of the Devonian.

The thickness of the whole Silurian in this part of the State given above as about 5,000 feet, is only an estimate. The true thickness it will be very difficult to determine, especially in the case of the most important member, the Knox Dolomite, since it is in great part made up, so far as surface materials are concerned, of loose fragments of chert in which the bedding planes are seldom to be seen. A greater part of the area of our valleys is held by this formation than by any other, and its importance is still further enhanced by the fact that it is the chief source of the brown iron ores of the State. Many of the noted big springs issue from this formation.

THE DEVONIAN.—The only representative in Alabama of this system of rocks, which in the States further north is of great thickness and importance, is a thin bed of *Black Shale*, averaging perhaps ten or fifteen feet, but being apparently absent altogether in some places. A few fossils have been found in it in the Valley of the Tennessee in North Alabama, which serve to fix its position as a member of the Devonian. The shale being soft and somewhat easily eroded, is usually covered and concealed by the debris of the adjacent rocks, so that it does not commonly come under notice even where it is present. It is of importance chiefly, perhaps, as being the source of some of our best known sulphur springs. The shale usually contains a large amount of pyrite in the form of nodules or kidney-shaped concretions, the decomposition of which supplies the sulphur of these springs. In North Alabama the thickness of the Black Shale may go up as high as 100 feet, but so extreme a thickness is rarely seen further south.

THE CARBONIFEROUS.—This we conveniently divide in Alabama into two parts, a lower, or *Sub-Carboniferous*, and an upper or coal bearing part, the true *Coal Measures*. The thickness of the latter is placed by Mr. Squire at 5525 feet, and the former at 1,200, making a total of between 6,000 and 7,000 feet.

Sub-Carboniferous.—Prof. Safford divides this formation in Tennessee into an *Upper or Calcareous* member, and a *Lower or Siliceous* one. This division will also apply equally well to that part of Alabama north of the Tennessee river, but to the south, and everywhere in the narrow anticlinal valleys of the State, this division will not suit, and we are compelled to make a different one. Like Prof. Safford, however, we make a two fold division, the *Fort Payne Chert* below, and the *Oxmoor Sandstone and Shales,* and the *Bangor Limestone* above, roughly corresponding to the divisions of Prof. Safford, with the differences below specified.

In the Tennessee Valley, the siliceous member of the Sub-Carboniferous consists of a great series of cherty limestones somewhat analogous to the Knox Dolomite, but with the lower part more cherty than the upper. This lower part gives rise to rather poor siliceous soils, and the region of its occurrence both in Alabama and Tennessee is known as the "Barrens"; the upper part of the Siliceous member is more calcareous and the soil derived from its disintegration is a red loam of more than ordinary fertility, well known in the Tennessee Valley as making the best farming lands of that section. Here again there is an analogy to the Knox Dolomite, which affords on the one hand rich red loam soils, and on the other poor cherty ridges.

The chert of the Sub-Carboniferous is in general very similar to that of the Knox Dolomite, but differs from it in being usually very highly fossiliferous, containing the casts or moulds of shells that have been leached or dissolved out.

This character of the Sub-Carboniferous chert, and the presence of the rhombohedral cavities in the chert of Knox Dolomite enable us in almost every case to distinguish between the two.

Now, in the anticlinal valleys south of the Tennessee river we find it impossible to carry out this two-fold division of the lower or Siliceous member of the Sub-Carboniferous, for the entire member shows, upon the surface at least, little else than chert, which appears in a mantle of angular fragments, covering usually one side of all our Red Mountain ridges. *a*

We have called this the *Fort Payne Chert*, and it is probably the representative of both the subdivisions of the lower Sub-Carboniferous or Siliceous group, of North Alabama and Tennessee, as long ago conjectured by Prof. Safford. Its thickness is not very great as compared with that of the upper member.

The Upper Calcareous member is variable in composition. In North Alabama it is chiefly a limestone called *Mountain Limestone*, from the fact that it forms the flanks of most of the mountains in that section that are capped with the Coal Measures. *b*

Within this limestone there is interbedded a layer of sandstone of variable thickness, perhaps 100 feet at a maximum in the Tennessee Valley, while the over and underlying limestones are many times that. As we come southward, the sandstone becomes more important, and the lower section of the limestone appears to give way to, or to be replaced by, a series of black shales closely resembling those of the Devonian but many times more massive. In many places in the anticlinal valleys, and especially the further south we go, the upper limestone also appears to be wanting or to be replaced by the shales and sandstones above named. The limestone which comes next below the Coal

a We have already adverted to the fact that these Red Mountain ridges are formed of the Clinton, the Black Shale and the Sub-Carboniferous chert, and the same structure has been mentioned by Safford as characterizing the Dye Stone ridges of Tennessee.

b The name, however, comes from Europe, where it appears in similar relations to the Coal Measures.

Measures is well exposed at many places as at Bangor, Blount Springs, and Trussville, where it is very extensively quarried for use as a fluxing material in the furnaces, as it is in part a very pure limestone, but south of the latitude of Birmingham it is very rarely seen, and in its stead we find the black shales mentioned. These shales are often interstratified with dark colored limestones and sometimes with tolerably pure limestones, but these are unimportant in thickness as compared with the shales and sandstones. The greater part of Shades Valley is based upon these sandstones and shales, though the limestone appears in several places.

The sandstone which in North Alabama lies between the two beds of Mountain Limestone, has a very close resemblance in texture and other characters to the lowermost rocks of the Coal Measures commonly called the Millstone grit, and it makes its appearance in that part of the State either as a bench along the sides of the Cumberland Mountain spurs, or else as the capping and protecting rock of a detached ridge separated from the Sand Mountain (Coal Measures), by a narrow valley of erosion. In the anticlinal valleys further south, this sandstone with the lithological characters above named, appears commonly as a distinct ridge running parallel to the escarpment of the Coal Measures, with a narrow valley of shales between. It appears to best advantage on one of the detached ridges above spoken of, near Tuscumbia, at the site of the old college town of Lagrange, and we have often used the name *Lagrange Sandstone* to designate it; but the name Lagrange has been used to denote an entirely different formation which has caused us to replace it by the name *Oxmoor*, where the rocks are also well exposed, and where the shales are more conspicuous than at Lagrange.

Coal Measures.—Of these rocks it does not seem necessary to speak in detail, since Mr. Squire has described the Coal Measures of the Cahaba Field, and since the measures of all the Alabama fields were probably once continuous, the description of the rocks of one will answer for all.

CRETACEOUS.—In the lower part of the area shown in the map our study of the distribution of the rocks of older formations is often very much hindered by the fact that they are more or less completely covered by superficial beds of sand and clay which have been spread over them after they had through the agencies above spoken of, been carved into topographic forms substantially the same as they now exhibit. The materials of this later formation are often distinguished by a purple or dark red color, the sands are mostly yellow, and show lines of cross-bedding, the gravels are unevenly distributed, and much less abundant than the sands. The clays as well as the sands with which they are interstratified, are more particularly characterized by the purple color mentioned, but there are many beds of the clay that are light gray and white. In a few places these clays are utilized for making refractory bricks, and the better grades of pottery, as at Woodstock, Bibbville, and Tuscaloosa. With careful selection and manipulation, there is hardly - doubt that these clays will be found suitable for all the uses to which the New Jersey clays are put, since they are essentially similar and belong to the very same geological formation. The formation contains a good deal of iron, which appears in the form of sandy and aluminous ores with 25 to 35 per cent. of metallic iron, usually scattered over the summits and along the slopes of the low hills of this region. The per cent. of iron is as a rule too low, and that of the silica too high to permit of these ores being used while we have such an abundance of ores of better grade.

POST TERTIARY.—Over the greater part of the State, except perhaps the extreme northeast, we find surface beds of very similar materials to those just described overlying the older formations. From about the limits marked on the map for the Tuscaloosa beds to the extreme border of the State towards the southwestward, we find these later beds occupying the surface, often to the extent of completely hiding the older rocks below, and forming the great bulk of the cultivated soils from the latitude of Tuscaloosa down. The distribution of these later beds within the limits of this map may be considered the same as that shown for the Tuscaloosa, and indeed where one is present the other is also in most cases, the Tuscaloosa below, the Orange Sand, as it has been called, above. Until a few years ago, they were universally confounded or at least not distinguished from each other, and the whole of these surface beds were thought to be Post-Tertiary, a confusion that very naturally followed from the great similarity not only of the material but of the mode of distribution, and the stratigraphy. In former reports we have called these Drift beds, but it seems bestto employ the name originally used by Dr. Hilgard to designate them, viz., Orange Sand.

In his report Mr. Squire speaks of the Drift beds which cover so much of the Coal Measures of the Cahaba Field in its lower part. These covering beds are in reality both Drift or Orange Sand, and Tuscaloosa.

In the coloring of the map it has not been attempted to show the Orange Sand, since its distribution is to all intents and purposes identical with that of the Tuscaloosa formation.

III. DISTRIBUTION OF THE ROCKS OF THE DIFFERENT GEOLOGICAL FORMATIONS IN THE VALLEYS BORDERING THE CAHABA COAL FIELD.

In the preceding pages we have endeavored to describe in a general way, the foldings, fractures, and displacements which the great rock masses of the Appalachian region have sustained through the action of the lateral pressure to which they have been subjected. This was done for the reason that, without some knowledge of the main types of geological structure prevailing in this region, it would be impossible to account for the present distribution and attitude of rocks of the different geological formations which appear in the two valleys which we shall attempt to describe.

We have already referred to the fact that with the flexing of the strata the crests of the arches, being lines of greatest strain were weakened, and fractured, and thus more easily wasted by erosion, and it is not surprising that, in process of time through the action of denuding forces, valleys should come to occupy the places once held by these arches. It is also plain that when the crests of these arches have been carried away by erosion, the remnants of the strata composing them will be exposed in the valleys in parallel bands, the oldest formation in the central part or axis of the valley, while on each side of this axis, and dipping or sloping away from it in opposite directions (anticlinal), will occur in regular succession, the newer formations up to the highest. Thus, beginning with the Coal Measures on, say, the northwest side of such a valley and crossing it towards the southeast, we should pass in succession over the strata, all dipping to the northwest, of the *Sub-Carboniferous Devonian* and *Silurian* to the *Cambrian*, which, as the lowest' of the geological series, would occupy the central area. Beyond this then would follow, on the other side of the axis of the anticlinal, the same formations, only in the reverse order, and dipping towards the southeast; thus *Silurian, Devonian, Sub-Carboniferous*, to the Measures of the Coal Field on that side.

Now, as a matter of fact, simple, symmetrical, anticlinal structure is rarely seen in any of our valleys, the nearest approach to it in the region here treated of being east of the Blount Mountain, and east of McAshan Monutain, but in both these cases the full series is lacking on one side of the anticlinal, by reason of a second fold or of a fault, as will be seen in the special description given further on.

As a rule we find a prevalence of southeasterly dips even on the northwestern side of the anticlinals. This could come about only by the overlap of the fold in that direction and the compressing together of strata so that they all dip the same way; or by an overlapped fold combined with a fault. In the first case we should have a repetition of the strata on each side of the central area, while in the other case only a part of the constituent strata of the anticlinal would appear on one side of the anticlinal, the rest being hidden under the overthrust measures of the other side. A study of the map will show that the last named order of things prevails in the great majority of cases.

Before going on to the special description of the valleys, it will be instructive to give a general section across the whole area of the map at a point where the structure is seen in its simplest form.

The accompanying diagram showing a cross section from

the Warrior to the Coosa Coal Field, through Birmingham, exhibits the main types of geological structure occurring in this part of the State, with the exception of those folds which show a prevailing dip in the northwest direction, of which mention has been made above, and which will be more particularly described in the proper place.

It must, however, be borne in mind that the diagram is not intended to give with absolute fidelity the section across the valley along a particular narrow line, but is rather intended to give the extremes occurring within somewhat widely separated limits. To illustrate: the red ore of the Clinton formation appears in Little Oak Mountain "i" in one or two places only, in the Cahaba Valley; and still less frequently, or rather in a much more fragmentary way, does it appear on the flint ridge "a" west of Birmingham; to the west of the fault beyond Opossum Valley we scarcely ever see so full a series as here shown of the beds between the fault and the Coal Measures in the vicinity of this cross section, though it appears further to the northeast towards Murphree's Valley. Keeping these things in mind, we shall find the diagram of service.

Beginning on the right hand of the diagram we see the Measures forming the northwestern border of the Coosa Coal Field overlooking with a steep face the valley to the northwest, the strata of the field dipping back to the southeast. Going thence to the northwest across the valley, we pass over the beds of the Sub-Carboniferous, Devonian, Trenton, Knox Dolomite, and Cambrian; all dipping southeast, and all forming the half of a fold or anticlinal uplift. But next beyond the Cambrian we come to the strata of the Cahaba Coal Field, with a vertical dip, and in immediate contact with the Cambrian; an association of strata which could come only from a break and sliding of the beds on one side of the break upon and over those on the other side. We see here that we have only the one side of a fold, or arch, and that a break has occurred along the crest of this fold, and the southeastern side has glided up over the northwestern side. We also observe that the beds of the Coal Measures adjacent to this break stand at a vertical angle, as a result of the break and the sliding up of the Cambrian beds. Beyond this point the strata of the Cahaba Field soon flatten down, and assume a dip to the southeast, these southeasterly dipping beds taken together with the vertical ones just mentioned, constituting a synclinal basin with its axis very near to the southeastern edge. The coal beds occurring in the vertical measures are undoubtedly the same as those occurring in the flatter measures just beyond, but we have the authority of Mr. Squire for saying that it is in most cases impossible to correlate the seams in the vertical measures with those that have not been so much disturbed. It is evident from this that the fault has broken up and displaced these coal seams so that they do not now occupy their relative positions in every case. As we cross the Cahaba Field we notice that the strata, with local exceptions, have a dip to the southeast, and the prevailing dip shows that the strata are gradually rising into another anticlinal fold which also includes all the underlying formations of Shades Valley, Red Mountain, and of the Birmingham Valley, as far west as the foot of the flint ridge "a" upon which is the cemetery. Here occurs another fault of the same nature as the one first described, except that the amount of the displacement is not by any means so great. At the eastern foot of this flint ridge, we find the strata standing in many places nearly vertical, as they do at the eastern edge of the Cahaba Coal Field. Along this line of fault the Cambrian of the valley lies in contact with the strata of the Knox Dolomite in most places, but an occasional bed of limestone and numerous fragments of red ore containing fossils which belong to the Clinton fauna, show that the Trenton and the Red Mountain or Clinton groups of the upper Silurian formation have not been entirely removed in the erosion of the valley.

Beyond the flint ridge just mentioned, we come, in going westward, again to the Cambrian strata, which, in a great measure, form the underlying beds of this second valley known as 'Possum Valley. Across it we come to a third fault which brings this Cambrian formation in contact with overlying beds, such as Trenton, Clinton, Sub-Carboniferous, and Coal Measures; for the fault does not by any means

SECTION N.W. AND S.E. FROM WARRIOR TO COOSA COAL FIELD IN THE VICINITY OF BIRMINGHAM
(nearly along line E.E of map)
showing structure of Jones' Valley–Cahaba Coal Field–Cahaba Valley and part of Coosa Coal Field

| Warrior Coal Field | Opossum Valley | Jones' Valley | Red Mt | Shades Valley | Cahaba Coal Field | Cahaba Valley | Coosa Coal Field |

Coal Measures
Bangor Sandstone and Shales | Sub-Carboniferous
Fossiliferous Chert Fort Payne
Black Shale, Devonian
Red Mountain Cluster
Trenton Limestone
Red Ridges | Knox Dolomite
Red Sands
Coosa Valley or Flatwoods shaly limestones — Cambrian

HORIZONTAL SCALE 2/3 OF INCH TO THE MILE OR
1 INCH = 7920 FEET
VERTICAL SCALE TWICE THE HORIZONTAL OR
1 INCH = 3520 FEET

a = North Highlands - Cemetery Ridge
b = Birmingham
c = South Highlands
d = Red Mountain
e = Shades Creek
f = Shades Mountain
g = Cahaba River
h = Anderson Mountain
i = Little Oak Mountain
j = Big Oak Mountain
k = Double Mountain

run strictly parallel to the strike or outcrop of the rocks, but runs in a sinuous line now into the very edge of the Coal Measures, now out into the valley so as to leave some of the strata underlying the Coal Measures between the Cambrian and the Warrior Field. As in the preceding cases, the beds immediately to the northwest of the fault usually stand at a high angle, sometimes vertical, while in places they have been pushed even past the vertical so as to be reversed.

On the northwestern side of the valley nearly its whole length, we find the first beds of the Warrior Coal Field, in this nearly vertical position, making a rock wall, through which the streams have cut their way at a few points, by deep and narrow gorges.

In cases where the strata bordering a fault are tilted up at these very high angles, it rarely happens that the full thickness of the beds concerned is present, but some are pinched out almost completely, others appear in full force, while still others are seen in exceptional thickness. Where beds of coal are among the strata, the thickness is nearly always found to be extremely variable, the beds thinning down to a few inches and thickening up to several feet within a distance of a few yards.

There is hardly a place along the western side of Jones' Valley where these irregularities are not to be seen. At North Birmingham, in one place, the Cambrian of the valley is, by the regular fault on that side, brought in contact with the upper part of the Knox Dolomite. Now we should expect to find beyond these beds of Knox Dolomite, first the Trenton, then the Clinton, the Black Shale, and Sub-Carboniferous, then the Measures of the Warrior Field; and very often such a succession of the beds does actually occur; but at the point named, the Knox Dolomite is in immediate contact on that side with the Coal Measures, the intervening strata above enumerated, having been pinched out or engulfed in a second fault.

CAHABA VALLEY.

This is the name given to the valley which separates the Coosa from the Cahaba Coal Field, and under this name it extends from near Odenville to Montevallo, but its continuation may be followed as far as Centerville. Like most of the valleys of similar nature in Alabama, it is complex, i. e., made up of one or more subordinated valleys with ridges between them. One of these valleys, lying between the chert ridge of the Knox Dolomite and the edge of the Cahaba Field, is known as 'Possum Valley; the other lying between the chert of Knox Dolomite and Little Oak Mountain is in the Cahaba Valley proper. In the Cahaba Valley (taken in its widest sense) there are the representatives of all the Paleozoic rocks above named, from Cambrian to the Coal Measures. Its two borders are made by the rocks of the Cahaba Field on the one side, and by those of the Coosa on the other, from its upper end near Odenville down to Siluria, while beyond that the Sub-Carboniferous beds of the Coosa Field make its southeastern border, since the Coal Measures of that field do not extend further south than the place named. Most of the strata in this valley dip towards the southeast at varying rates, from which it would be reasonably inferred that its structure is that of an anti-clinal fold closely compressed and pushed over towards the northwest, or of an anticlinal fold and thrust fault combined. In the former case, we should, in crossing the valley, pass over the strata from Coal Measures of the Coosa Field to the Cambrian in succession, beyond which should follow the same formations again, only in reverse order, to the Measures of the Cahaba Coal Field.

The diagram already referred to, as well as the examination of the map will show, that the whole valley is made by one-half of a fold only, and the succession of the rocks from the Coosa Field is as follows : Coal Measures, Sub-Carboniferous, Black Shale, Trenton, Chert ridge and red lands of the Knox Dolomite and Cambrian, immediately following which are the Measures of the Cahaba Field, a great fault intervening between the Cambrian and Coal Measures. By this fault the Cambrian strata on the southeast side have been pushed up and over the upturned edges of the Cambrian, Silurian, Devonian, and Sub-Carboniferous on the northwest side, into direct contact with the upper measures of the Cahaba Field, a displacement of perhaps more than 10,000 feet vertical, and greater than that of any other fault known to us in Alabama. Northwest of this fault, as we have already intimated, all the formations between the Cambrian and the Coal Measures are below the surface being engulfed in the fault. The beds of the Coal Measures next to the fault are very highly inclined, standing mostly nearly in vertical position while sometimes they have been pushed over beyond the perpendicular. The narrow belt of vertical measures borders the Cahaba Field along its entire southeastern and southern boundary. We should naturally expect the strata in these vertical measures to correspond with those that have not been disturbed further in towards the center of the field, since they are only the upturned edges of the same beds; but Mr. Squire has generally been unable to identify the vertical coal seams, for the reason that in the faulting the strata have been so crushed and displaced that the seams no longer retain their characteristic qualities, thickness, relative position, etc., some of the measures having been pinched out, and others having been correspondinly thickend up. This, in general terms, is the structure of the valley from its northern limit to Siluria, and even down to Montevallo.

In more detail, its topographical and geological features are as follows: The southeastern rim of the valley is made by the high escarpment of the Millstone grit of the Coosa Field known as Big Oak Mountain. This ledge of rock dips southeast under the Coal Measures of the Coosa Field, but is brought to the surface again in the Double Mountains, by. a fault that extends through the lower part of the field.

Between Big Oak Mountain and Little Oak, which is formed by the chert of the Sub-Carboniferous formation, there is a valley of varying width formed by the Oxmoor shales of the same formation. The sandstones which accompany these shales, form one or more small ridges between Little and Big Oak Mountains, and in some parts of the valley this sandstone extends a good way up the face of Big Oak, and then the Millstone grit forms only the capping of the mountain. Little Oak is the counterpart of the Red Mountain, but the Clinton strata appear to be entirely wanting except in two or three places shown on the map.

To the northwestward of Little Oak Mountain comes a

valley underlaid by the Trenton or Pelham limestone, which is here remarkably well developed, and extensively quarried to supply the lime kilns at Siluria and Longview. This lime is the well known Shelby lime.

Beyond this Trenton limestone valley, which is the Cahaba Valley proper, towards the northwest comes a high ridge formed of the chert of Knox Dolomite, known in its different parts as New Hope Mountain, Mill Ridge, Pine Ridge, and Anderson Mountain.

Next follows a valley based upon calcareous parts of the Knox Dolomite and the variegated shales of the Cambrian. This has the name of 'Possum Valley, and as we approach its northwestern edge we find the strata gradually assuming a steeper dip up to the edge of the great fault spoken of above, and beyond this fault are the vertical measures of the Cahaba Field. It may be noticed here as almost everywhere else that the dip of the strata on the southeast side of one of these thrust fault is usually considerably less than on the northwestern side, where they frequently stand nearly vertical. This is in conformity with the law of structure that prevails through the whole Appalachian region, viz., the steeper dip is on the northwestern side of the folds and faults, except where there has been an undershoving of the strata, as is the case in Murphree's Valley. Beyond the fault, the measures very rapidly flatten down to a moderate rate of dip, which is mostly towards the southeast, showing that taken as a whole this upper part of the Cahaba Field is a synclinal basin, the axis of which is very close to its southeastern boundary. The southeastern half of this synclinal is partly engulfed in the great fault, for there is usually not room enough between the axis of the synclinal and the boundary fault for the whole thickness of the Measures to come in, even in vertical position.

Southward of the latitude of Siluria the coal-bearing measures of the Coosa Field give out, but the underlying Sub-Carboniferous strata continue as far as the limits of this map, and beyond even, till they are completely hidden below the Cretaceous beds of the Tuscaloosa formation. This southward prolongation of the Coosa Field is made chiefly by the Oxmoor shales and sandstones, which, especially the former, attain very considerable thickness and underlie a wide area. Beds of limestone are very generally interstratified·with these shales.

By referring to the map it will be seen that the Cahaba Valley in its upper part runs nearly northeast and southwest, but below Helena it turns nearly southward to Montevallo, while beyond the latter point the edge of the Cahaba Field turns nearly west, as does also to some extent the Sub-Carboniferous border of the valley on the other side. These changes in the direction of the folds, bring about complications of the structure, as may be seen in the formation of a great number of subordinate basins in this part of the Cahaba Field; in the faulted and overturned measures west of Montevallo; in the faulting and duplication of the Cambrian and Silurian strata in the valley between Montevallo and Centerville; in the formation of a synclinal of Trenton, Red Mountain, Devonian and Sub-Carboniferous strata in the vicinity of Pratt's Ferry.

The uppermost beds exposed in this synclinal are the Oxmoor shales which are seen in the basin of the "Mountain," which, beginning a mile below Pratt's Ferry, extends in V shape for several miles towards the southwest. The point of this mountain or apex of the V overhangs the river. The dip of the northwest side is moderate towards the southeast, while the strata on the southeast side are almost vertical, following the usual law. The crest of the mountain is formed by the Fort Payne Sub-Carboniferous chert, while upon its flanks are to be seen the underlying rocks down to the Trenton. Below the chert there are sandstones and shales that we have referred to the Clinton or Red Mountain, though we have no fossils nor any of the red iron ore to determine the matter. *a*

The Trenton rocks underlying this Sub-Carboniferous basin emerge from below it, both towards the northeast and to the southwest, but more rapidly in the last named direction. Towards the northeast the Trenton belt may be fol-

---

a Some red ore which occurs a mile or two to the northwest of the "Mountain," appears to belong to a bed lying between the uppermost of the Knox Dolomite and the lowermost of the Trenton. At least such is its position at one locality where all these beds may be clearly made out, and their ages distinguished by the fossils which they hold.

lowed for a considerable distance, gradually merging into a fault which cuts into it in the direction of Montevallo.

One of the most important results of this disposition of the rocks is to bring to the surface a great area of Trenton limestone with very gentle dip, except at the southeastern edge of the basin, all along the river for several miles each way from Pratt's Ferry. Much of this limestone is of very great purity, and is capable of receiving a fine polish, and it will undoubtedly very soon come into use for building and ornamental purposes.

That part of the valley below Montevallo differs slightly in structure from the upper part. Starting at the Sub-Carboniferous measures which here form its southeastern border we pass over a valley of Trenton limestone, then over a wide area of Knox Dolomite, three or four miles, chiefly cherty lands, into a belt about a mile wide of the Cambrian shales of the Montevallo type. Then comes a fault by which the Knox Dolomite is again brought to the surface. This narrow belt of the Knox Dolomite runs out entirely in township 24, range 11, east, but at the base of the map in township 24, range 10, it is perhaps half a mile wide. The southeastern edge of this belt of Knox Dolomite has a nearly vertical position, and, together with part of the Trenton, forms the edge of the synclinal above named.

In the upper part of the valley down to about the upper line of township 22, the edge of the coal field runs approximately parallel to the strike of the rocks exposed in the valley, but below the point named this is not the case, for the strata of the Montevallo shales that are in contact with the vertical measures of the Cahaba Field through township 22, and the upper part of township 24, have a strike nearly northeast and southwest while the edge of the coal field runs nearly north and south; the Cambrian strata appear to run up against the Coal Measures at an acute angle; and while the border of the Cahaba Field above Montevallo runs nearly north and south, changing abruptly at Montevallo to nearly east and west, the strike of the older rocks remains approximately the same, i. e., nearly northeast and southwest. At the apex of the right angle formed in this change of direction, a little southwest of Montevallo, near Thompson's

COAL SEAM UNDER CAMBRIAN LIMESTONE.
NEAR THOMPSON'S MILL, SHELBY CO., ALA.

Mill on Shoal Creek, there is one of the most interesting sections known to me. Here may be seen a bed of coal three or four feet in thickness, in nearly horizontal position, with the shaly limestones of the Montevallo series resting directly upon it. The accomyanying view from a photograph shows this very clearly. Mr. Squire has shown that the Coal Measures along this part of the field have been overturned, and the bottom fireclay is in every case on top of the seam. In the faulting, therefore, not only has a large strip of the Coal Measures been pushed over, but the Cambrian strata have been slipped up and over these reversed beds.

The map does not show very clearly the manner in which the Cambrian passes around the apex of this angle of Coal Measures, for in reality these older measures seem to lap up upon the angle of the Coal Field in a series of great parallel waves like breakers upon an exposed point of the shore. These waves do not accommodate themselves to the turn in the boundary of the Coal Field by bending round, as might be inferred from the arrangement of the colors on the map, but they keep their original direction, (northeast and southwest), on the two sides of the salient angle, just as waves pass an obstruction.

All along the Cahaba Valley and its extension southward and southwestward of Montevallo, the area formed by the Knox Dolomite is characterized by the occurrence of beds of brown iron ore or limonite that in many places are destined to be af great economic value.

For lack of means of transportation, only one furnace has up to the present time been built to utilize these ores.

Coosa Coal Field.

The structure of the Coosa Coal Field does not at this time particularly concern us, but the portion of it included in the map shows that it is divided into two parts by a fault which brings up some of the Sub-Carboniferous shales between the two. This belt of shales varies in width from half a mile upwards, and the amount of displacement is not very great, since it extends only from the lower part of the shales up to the Millstone grit. Mr. McCalley's report will give a tolerably full account of the structure of this field.

JONES' AND ROUP'S VALLEY.

An inspection of the map will show that the long valley separating the Cahaba from the Warrior Coal Field, is much more complicated in its structure than the valley between the Coosa and Cahaba Fields.

Like the Cahaba Valley, this has essentially an anticlinal structure, and like that valley, this structure is somewhat masked by faulting; but in addition to this we can trace out in every part of this valley, two anticlinal folds separated by a synclinal. Almost everywhere in the valley the anticlinal folds have been pushed over towards the northwest in accordance with the general law of Appalachian structure, and the axes of the folds are close to their northwestern edges. In the synclinal we find its axis near the southeastern edge, as is the case in the upper part of the Cahaba Coal Field.

There are two classes of exceptions to this general plan of structure noticed in Jones' Valley. First, where the anticlinal is nearly symmetrical, and the strata on the two sides of it dip in opposite directions at approximately the same angle. One instance of this may be seen in the valley between McAshan Mountain and East Red Mountain, and another in the upper part of the valley west of Springville, in Clayton's Cove and northeastward; both of which will be more particularly described in another place. In these cases also the crest of the anticlinal is unbroken, while everywhere else the crests are marked by thrust faults.

The second class of exception to the general plan of structure is seen in those cases where the strata dip towards the northwest, and the fault is found along the southeastern border of the arches, making what we have spoken of above as a reversed thrust fault. Two well marked instances of this class of exception occur in Jones' Valley; one being west of McAshan Mountain, the other being in the northeastern part of the region of the present map; but the most important instance is in Murphree's Valley. *a* The case of

---

a To Mr. A. A. Gibson belongs the credit of first calling attention to this type of structure in Alabama. In his report on Murphree's Valley, now in manuscript and soon to be published, will be found full details of the typical locality.

McAshan Mountain will be considered further on, but we may conveniently now describe the occurrence in township 15, range 1, east, along the northeastern border of the valley. Here, not far from the line of the A. G. S. R. R., there is a fault along which on the northwestern side the strata of the Knox Dolomite with moderate northwesterly dip, are in contact with the strata of the Sub-Carboniferous of the Cahaba Field on the southeast, with nearly vertical position or with very high southeasterly dip.

This is the reverse of the usual order of things, and may be explained as already shown upon the supposition that the fold, instead of having had its crest *pushed over* towards the northwest, has had the trough *shoved under* from the southeast side. At the lower end of this fold in the northwest corner of township 16, range 1, east, its anticlinal character is more apparent. Beyond this fold to the northwest, we see a synclinal with Sub-Carboniferous chert and Oxmoor shales, as the uppermost beds, and this is followed in the same direction by a simple anticlinal and then by the synclinal of Blount Mountain (Coal Measures), northeast part of township 15, range 1, west, and beyond that the anticlinal of Murphree's Valley not shown in the limits of this map. Southwest from the end of this Blount Mountain synclinal of the Coal Measures we see the underlying beds gradually coming to the surface in the order, Sub-Carboniferous, Devonian, Red Mountain, Trenton, and Knox Dolomite. The upper formations involved in this synclinal from the Sub-Carboniferous to the Trenton, do not extend many miles below the end of the Blount Mountain, but the synclinal of the Knox Dolomite may be followed down nearly to Bessemer. It makes all that ridgy land between Chalkville and Hagood's Cross Roads, the flint ridge of the North Highlands about Birmingham, and its continuation down to the old Smith place near Bessemer. Indeed, with certain modifications it may be followed almost the entire length of this map.

From Trussville down to the lower end of Jefferson county, in the southeast part of township 20, range 5, west, the feature that perhaps most strikes the eye is the wide valley based on the upper rocks of the Sub-Carboniferous, viz., the Oxmoor shales and sandstones, lying between the edge of the Cahaba Field and the East Red Mountain. This valley is in great part drained by Shades Creek and is known as Shades Valley. Its abnormal width is due to the undulations in the strata, since the Sub-Carboniferous beds are no thicker here than in other parts of this valley where the width is much less. These undulations are accompanied by faults in some parts of Shades Valley, as for instance between Oxmoor and Grace's Gap, but these displacements have not yet been traced out with sufficient detail to permit of their being properly mapped.

Shades Valley is diversified by long ridges formed by the sandstones of the formation, and it is usual to find a very distinct and persistent ridge near the western edge of the valley formed by sandstones that occur near the base of the formation. Limestones occur in these shales, as has been already noted, and in one place near Oxmoor this rock has been quarried.

The next following topographic feature to the northwest of Shades Valley, and by far the most important one in the region, from an economic standpoint, is the Red Mountain. In the lower part of the area shown on the map, i. e., below the crossing of the Cahaba Coal Company's railroad, the Red Mountain does not form a conspicuous topographic feature, as it is rather low and in many places covered by the sands and other beds of the Tuscaloosa formation. Above the point named, it begins to assume, at least in places, the dimensions of a mountain, and so it continues with constantly increasing height and importance almost to the upper limit of the map. I shall not attempt here to speak in detail of the variations observed in the strata of Red Mountain, nor to give sections across it, since the report of Mr. McCalley, soon to be published, will fully treat of this part of the subject. Most of the mines at present in operation in the Red Mountain are found between Spark's Gap and Trussville, the greatest thickness of ore, about twenty feet, being about the middle part of this stretch of the mountain. Above Gate City, Red Mountain turns somewhat away from the edge of the Cahaba Field, and the reversed anticlinal above spoken of, comes in between the two, and by this a synclinal is also formed in the Red Mountain strata. The Red Mountain has everywhere along its eastern flank a covering of the chert of the Sub-Carboniferous, and the Black Shale, which comes between the Clinton and the Sub-Carboniferous, while not always to be seen on account of its being very thin and easily eroded, is no doubt present in the majority of cases. On the western face of the Red Mountain, the Trenton limestone may always be seen, sometimes near the base of the mountain, sometimes nearer the top, according to locality, and this rock is extensively quarried, notably at Gate City, where the limestone extends up to the very top of the mountain, and the Clinton strata are all on the eastern flank of the same. This varying position of the Trenton is due to local causes, among which the occurrence of undulations running across the valley is perhaps the most effective.

Next to the Red Mountain with its constituent formations, follows the Knox Dolomite, making first a belt of ridgy lands, seen in the South Highlands, and then the redlands with their gentle undulations and characteristic soils, as may be seen near Elyton and in some parts of the city of Birmingham itself. It is rarely that the strata of the Knox Dolomite appear in their original form so that their dip may be clearly recognized. Usually the formation is represented by great accumulations of loose fragments of chert, or by the red loams in which bedded rocks are rarely found. Loose angular fragments of chert imbedded in the red soil are however very common and characteristic. This eastermost belt of the Knox Dolomite rocks presents no special features. In the lower part of the map it is in great measure covered by the Tuscaloosa sands and clays, though cropping out in spots over a pretty wide area here. On account of the covering of these surface materials it has thus far been impossible to make out with certainty the structure of all this lower part of the map.

In the upper part of the region covered by the map, we find a second wide and apparently continuous outcropping of the Knox Dolomite, I mean above Eastlake, up to the end of the Blount Mountain. This is due, as may have already been inferred by the reader of the preceding paragraphs, to the Blount Mountain synclinal above spoken of, and to the fact that the fault which borders this synclinal on the eastern side extends with constantly diminishing amount of displacement, only a short distance beyond Eastlake, where it gradually passes into the unbroken or unfaulted anticlinal of Clayton's Cove. In this way the Knox Dolomite of both anticlinal and synclinal are brought into juxtaposition, while further to the south, where the amount of displacement in the fault is greater, the two are separated by the belt of Cambrian Shales presently to be spoken of.

Next to the Knox Dolomite, going still across the valley, we come to the Cambrian formation, here represented by the Coosa Shales, a series of thin-bedded limestones with clay partings that make level, flat, badly drained lands with heavy impervious clay soils, commonly known as "Flatwoods." The flatwoods limestones are usually very much folded and contorted, and stand often nearly vertical, for which reason they were spoken of by Prof. Tuomey as the vertical limestones of the valley. We usually see the upturned edges of these limestone bands outcropping in the flatwoods in parallel rows, sometimes running without serious breaks for long distances. At McCalla Station, Bessemer, Powderly, and in parts of Birmingham, this limestone may be seen and easily recognized. These are the lowest in a geological sense, of the rocks brought up by the anticlinals and faults in our valleys, and are the oldest of the rocks of Alabama about whose age we can be perfectly sure. In a regular symmetrical anticlinal, in which these Cambrian strata were exposed by erosion, they would, as a matter of course, occupy the central area, and this is in reality the case in that part of the valley between McCalla Station and Tannehill; but in the far more common case, where the anticlinal is pushed over to the northwest and the steeper slope occurs on that side, and still more plainly, where a break occurs along the crest of the anticlinal and the strata on the southeastern side are slipped up over those on the northwestern, the Cambrian strata are to be found no longer in the geographical center of the valley, but far over on its northwestern side.

The Cambrian belt above described, thus marks the limit

of the first anticlinal of the valley, and adjacent to it towards the west but separated from it by a fault, is the flint ridge of the North Highlands, (Knox Dolomite). As we have already said this ridge is in structure a synclinal, with the axis close to the southeastern border, and with most of the strata on that side of the axis overridden and concealed by the Cambrian of the other side of the fault, a case exactly analogous to what we have seen along the southeastern border of the Cahaba Coal Field. In the fault above spoken of the adjacent halves of the anticlinal of Jones' Valley and of the synclinal of the flint ridge, are engulfed more or less completely, though we commonly find along the eastern face of the flint ridge a narrow belt of vertical or nearly vertical rocks which belong to the synclinal, and are the only remnants of its eastern half. As is the case on the corresponding side of the Cahaba Field, these vertical measures very rapidly flatten down and begin to rise on the other dip, so that the center or axis of the anticlinal is very close to this eastern edge.

As the name indicates, this ridge is formed mostly of the flint or chert of the Knox Dolomite, but there may be found at intervals between East Lake and Bessemer, traces of the rocks of other overlying formations, Trenton and Clinton, showing that these were also involved in the foldings, but have in great measure been removed by denudation.

Upon this flint ridge at several points, and beyond Bessemer in the Salem Hills, we see great masses of a peculiar rock, made up of angular fragments of the chert of Knox Dolomite cemented together into a firm and compact rock. This breccia is at the top of the Knox Dolomite, or perhaps it would be better to say, at the base of the next higher series, the Trenton, since it is made of fragments of the Knox Dolomite, and must therefore be younger.

This flint ridge is a marked feature of Jones' Valley, and extends without serious break from near Village Creek at Birmingham, to Valley Creek near Bessemer. Northeast of the former creek it is seen again, and southwest of Valley Creek it appears in the Salem Hills. At the two places mentioned the ridge is cut by the creeks, down through a good part of the chert of the Knox Dolomite into the red lands of the same formation, and the continuity of the ridge is interrupted. We have already intimated that in a modified form the synclinal of Blount Mountain is the continuation of this.

Going northwest beyond the flint ridge we pass over the strata of the red lands of the Knox Dolomite, then over a second belt of Cambrian, all dipping back below the ridge, and rising to the northwest into the second anticlinal, here called 'Possum Valley. The summit of this anticlinal, like that of Jones' Valley, marks the line of another thrust fault similar to that of the flint ridge, though much more irregular in its course, for while, along the border of the flint ridge, the fault brings the Cambrian as a rule in contact with the chert of the Knox Dolomite, with here and there an exception where it is brought up against Trenton and Clinton, in this second fault the Cambrian is brought up in contact with Knox Chert, with Trenton, with Clinton, with Sub-Carboniferous, and even with the measures of the Warrior Field. This fault hence shows a much greater variation in the amount of displacement than the one first named and described. This may be made clearer by reference to the section above referred to, and to the map. This fault runs along nearly parallel to the line of the Birmingham Mineral Railroad above Boyle's, up into Murphree's Valley. Above the line of the South & North Alabama Railroad, it will be seen that the fault is at some distance from the edge of the Warrior Field, and that strips of the following formations intervene between the two, viz., Knox Dolomite, Trenton, Clinton and Sub-Carboniferous, and that the fault passes from the western side of 'Possum Valley across to the eastern side of Murphree's Valley. As we approach Boyle's Gap, the width of the belt of intervening measures decreases, some of the formations seem to be pinched out completely others seem to be partly cut out, and none of them retain their full characters. The diagram (cross-section) shows the whole series from the Knox Dolomite up to the Sub-Carboniferous as intervening between the fault and the edge of the Warrior Field, which is in reality the case in some places, but we need only to examine the map to see how the fault runs irregularly along the border of the Warrior Field, now lapping up in contact with the rocks of the Coal Measures, now trending further out into the valley, leaving the upturned edges of the whole series from Knox Dolomite up, between.

West of Bessemer we see a rather complicated spot where the Red Mountain rocks attain a considerable development, which will be understood better by a study of the map than by any description in words.

As may be inferred from the map, the Red Mountain on this western side of the valley is rather fragmentary, and of little value as compared with the same formation on the other side, east of Birmingham. Above Boyle's Gap it becomes more regular and of greater economic importance. It need hardly be repeated that the strata of all the formations to the west of this second fault, stand at very high angles, often being perpendicular, and at times being pushed over past the vertical so as to dip back towards the southeast. The millstone grit of the Warrior Field may nearly always be seen as a ledge of nearly vertical rocks forming, most the whole length of the valley, a wall, beyond which we come in a few hundred yards to almost horizontal measures, showing that the disturbance affects to any great degree, only the extreme edge of the field. Parallel with this rock wall of the Millstone grit, we usually find another wall of vertical rocks, with a narrow valley intervening. This wall is formed by the Sub-Carboniferous sandstone of the Oxmoor series. The line of the fault may easily be traced by the ledges of vertical or nearly vertical rocks that lie to the northwest of it. Such, then, is the structure of the valley in all the upper half of the map, or above the latitude of Bessemer. Below that there are some important variations which have in part been referred to.

The variations from the above named structure are to be seen in the area through which the McAshan Mountains extends. This mountain is a Red Mountain ridge composed of the three formations, Clinton, Black Shale, and Sub-Carboniferous chert, with Trenton limestone on its eastern face. Beyond this mountain and across a fault, we find a repetition of the same beds, a second Red Mountain, in its normal place as regards the Warrior Coal Field.

Although so entirely different in topography and in general appearance, this part of the valley is itself also formed by a double anticlinal with synclinal between, as may be seen from the following description; proceeding from the eastern Red Mountain near McCalla Station towards the northwest across the valley, we pass first over a regular symmetrical anticlinal, the central line of which, marked by the outcrop of the belt of Cambrian rocks, is near the center of the valley, and is the formation upon which the Alabama Great Southern Railroad track is laid, from Tannehill up. McAshan Mountains is the counterpart of the eastern Red Mountain on the other side of the anticlinal, its strata dipping to the northwest as the beds of the eastern mountain dip to the southeast. On the other side of McAshan, however, we come to the fault mentioned, and the beds of the McAshan appear to dip northwest under the Knox Dolomite on the other side of the fault, showing that we have here again an instance of thrust fault in which the strata on the southeast side have been *shoved under* those on the northwest side.

Southwest of the end of the McAshan Mountain we see again a recurrence to the usual type of structure in this valley, viz., an overlap of the strata on the southeast side upon those to the northwest of the fault. As we have said, however, the geological structure in this lower part of the region of the map is not always to be clearly made out, for the reason that it is not possible to trace out the outcrops of the different formations because of the great mass of overlying and more recent beds of the Cretaceous. The central part of the valley in this latitude is so generally covered by these beds that we can only indicate here and there the points where the underlying rocks are uncovered.

In the vicinity of Woodstock there appear to be two areas of Cambrian rocks, the one at the station itself, where the shaly limestones have been exposed in the cut made by the Cahaba Coal Company for their railroad, and the other a mile or two to the north, along the line of the Birmingham Mineral Railroad, just beyond the Edwards ore banks. The region between the two, so far as we are in condition to judge, is occupied by Knox Dolomite.

It may be that the structure here is similar to that of the valley about Birmingham. In a cut on the Birmingham Mineral Railroad, just beyond the Edwards ore banks, the Cambrian limestone and shales have been laid bare, and exhibit one of the most perfect examples of the contortions and foldings into which it is possible to throw solid rocks. The limestones have been pressed together into a number of close folds, as perfectly and completely as one could do it with a bundle of sheets of paper. The edges of these folded limestone layers are seen in zig-zag lines all along on both sides of a cut of forty or fifty feet in length. These layers of limestone are quite pure and have been used in making lime which slakes very well, showing that it is of very good quality. Now, while at the base of the cut and for ten feet or so above the level of the track, the limestone is quite fresh, and unweathered, it passes very suddenly into a yellowish stratified clay in which may be followed perfectly all the lines of folding of the limestone itself, as if the upper part of the limestone, near the outcrop, and where long subjected to the action of the atmospheric agencies, had been converted into the clayey matter. If the limestone were impure and charged with clayey material, we might suppose that the calcareous matter was leached out and the aluminous part left, but the limestone is pure enough to afford good, thoroughly slaking lime, so that the whole appearance is as though the limestone had been removed by leaching agencies, and its place taken by a sandy clay. We should in any case expect to find a gradual transition from the one kind of material to the other, but as I have said, the change is rather abrupt.

The strata of the Red Mountain may be followed with some interruption from opposite Woodstock down to Vance's on the west side of the valley.

The fault which occurs on this side of the valley appears to run in and out approximately parallel to the edge of the Coal Field, now leaving a pretty full series of strata between it and the Coal Field, now lapping up almost upon the beds of the latter, by pinching out or engulfing the intermediate formations.

West of Vance's we see a narrow anticlinal fold which

runs a short distance up into the Coal Field and separates a small strip of synclinal structure from the main body of the Coal Field. This synclinal extends southwestward as far the limits of the map, with Sub-Carboniferous and Clinton rocks, the Coal Measures ending at about the latitude of Vance's station. The Clinton strata of this synclinal are much broken up and appear to be pinched out in places. The red ore occurs in the vicinity of Vance's, at one or two points southwest of the station, and in the railroad cut two miles west of the station. Further to the southwest than the points named, the Clinton is represented by sandstones and conglomerates alone, and the red ore seems to be wanting.

The anticlinal fold above spoken of is faulted near its central line, and the rim of the Clinton rocks which would normally run along the western side of the anticlinal has been cut out by the fault with the exception of a small remnant seen in the railroad cut above mentioned. By the fault a strip of Knox Dolomite has cut out about half of the anticlinal as shown on the map. To the northeast of the railroad the anticlinal is occupied only by the Oxmoor Shales of the Sub-Carboniferous. In addition to the great fault above noticed there is a smaller one which shows in the railroad cut to the west of the trestle over the branch of Hurricane Creek. This structure will be more easily understood from a study of the map than from the reading of a description. The superficial beds of the Tuscaloosa formation overlying all the older rocks makes it extremely difficult, and in some cases impossible, to determine with certainty the structure of the lower part of the valley south of Vance's.

INDEX.

	PAGE.
Acton Basin—Area	39
Boundary	39
Estimate of coal in	46
Sections of	44
Situation	39
Structure	39
Variations of dip of measures of	45
Action Seam—Acton basin, section of	43
Air Shaft Seam Daily creek basin	108
Lolley basin	87
Montevallo basin	92-94
Analyses Ash—Mammoth seam, Henryellen basin	38
Gholson seam, Dailey creek basin	110
Coals—Conglomerate seam, Eureka basin	72
Helena seam, Helena basin	59
Lemley seam, Overturned Measures	102
Little Pittsburgh, Eureka basin	72
Mammoth seam, Henryellen basin	31-32
Montevallo seam, Montevallo basin	94
Moyle seam, Eureka basin	72
Thompson seam, Eureka basin	72
Underwood seam, Blocton basin	115
Wadsworth seam, Eureka basin	73
Helena basin	60
Woodstock seam, Blocton basin	115
Cokes—Mammoth seam, Henryellen basin	38
Wadsworth seam, Cahaba basin	65
Woodstock seam, Blocton basin	116
Anthracite system of mining	118-119
Bangor limestone	155
Basins in Cahaba field	11
Acton	39
Blocton	111
Cahaba	61
Dailey creek	103
Dry creek	74
Eureka	68
Gould	78
Helena	47
Henryellen	20
Lolley	83

INDEX.

BASINS IN CAHABA FIELD—Continued.

Montevallo	90
Overturned Measures	95
Beebe seam in Overturned Measures	96, 99, 100
Beech Tree seam, Blocton basin	114
Dailey creek basin	107
Big Falls, Lolley basin	89
Big Vein (seam), Dailey creek basin	106
Birmingham Brecc.a	152
Black Fireclay Seam—Dailey creek basin	108
Lolley basin (section of)	88
Montevallo basin	92-94
Black Shale Formations	154
Black Shale or Gholson Seam—Dry creek basin	76
Eureka basin	69-70
Helena basin, section	53
analyses of coke from	59
Blocton Basin—Area	113
Boundaries	111
Dip of Measures	116
Drainage	112
Estimate of coal in	113
Faults	113
Synclinals and anticlinals in	116
Roads in	112
Topography	112
Brock Seam—Cahaba basin	63
Buck Seam—Dry creek basin	76
Eureka basin	69-70
Helena basin (section of)	52
Cahaba Basin—Area	62
Boundaries	51
Drainage	62
Estimate of thickness of measures	62
General section across	62
Roads	61
Situation	61
Topography	62
Varying rate of dip	65
Cahaba Field—Aggregate thickness of measures of	14
Amount of coal in	13
Area	13
Basins of	13
Conglomerates at top of measures of	4
Counties in which measures occur	17
Division of coals into four groups	14
Drainage	6-7
Faults in	15
General description	3
History of mining in	18

CAHABA FIELD—Continued.
Limestone ledge in	4
Overturned measures of.	15
Rate of dip of measures	17
Resemblance of measures of to those of Arkansas and Indian Territory	5
Roads	10, 11, 12
Sections illustrating structure of	13-14
Small amount of sulphur in coals	5
Cahaba Field—Similarity of measures to those of Warrior Field	3
Thickness of measures	5
Topography	6-7-8 9
Cahaba Valley—General description	163
Geological and structural details	165
Cannel Seam in Overturned Measures	97-99
Carboniferous Formation—Subdivisions of	155
Choccolocco Shales	149
Clark Seam Dailey Creek basin	106-107-109
Clean Coal Seam—Dailey Creek basin	107
Clinton Formation described	153
Coal Measures of the three Alabama fields once continuous	157
Coke Oven Seam—Helena basin	51
Cahaba basin	65
Coke Seam—Dailey Creek basin	106
Blocton basin	114
Combination Method of Mining	122-123
Conglomerate Seam—Analysis of coal	72
Dailey Creek basin	108
Eureka seam	69-70
Helena basin	55
Henryellen basin, identical with Thompson seam and Underwood seam	27
Lolley basin	85
Cooper Seam in Overturned Measures	96-98-99-100
Coosa Coal Field	169
Cretaceous Formation	157
Cubical Vein Seam—Overturned Measures	99
Dailey Creek Basin—Area	105
Boundaries	103
Dip of measures	108
Drainage	104
Estimate of coal in	105
First mining in	109
Roads	104
Topography	104
Devonian Formation	154
Dodd Seam in Overturned Measures	96-97-99
Drift Formation or Orange Sand	159
Dry Creek Basin--Area	75
Boundaries	74
Dip of measures	76

DRY CREEK BASIN—Continued.
    Drainage... 74
    Estimate of coal................................ 75
    Future importance of........................... 76
    Roads in.. 75
    Topography..................................... 75
Eureka Basin—Area...................................... 69
    Boundaries....................................... 68
    Drainage.. 68
    Estimate of coal in.............................. 69
    Method of working seams in..................... 71
    Strike of measures.............................. 69
    Topography..................................... 68
    Varying rate of dip in........................... 70
Eureka Company's Test Slope—Section of, in Acton basin.......... 43
Faults—Amount of displacement 16-17
    Boundary fault of Cahaba Field.................. 15
    Difference in angle of dip on two sides of....... 16-17
    Interior faults of Cahaba Field.................. 15
    Reversed thrust faults.......................... 142
    Thrust faults................................... 142
Figh Seam—Overturned Measures......................... 99
Five Group Seam—Acton basin........................... 45
Folds in strata in Valley region.......................... 140-141
Fort Payne Chert - Sub-Carboniferous 155
Gholson or Woodstock Seam—Blocton basin............... 114
    Dailey Creek Basin, sections and analyses..... 106-107-109-110
    Lolley basin........................ 85
Gould Basin—Area...................................... 79
    Boundaries....................................... 78
    Dip of measures................................ 82
    Drainage.. 79
    Estimate of amount of coal...................... 79
    Roads... 80
    Topography..................................... 79
Gould Seam—Acton basin................................ 44
    Blocton basin................................... 114
    Cahaba basin 63-65-66
    Gould basin.................................... 80-81
Half Yard Seam—Dailey Creek basin 106
Harkness Seam – Acton basin............................ 45
    Helena basin 64
Helena basin—Area..................................... 59
    Boundaries...................................... 47
    Disturbances in measures of.................... 48
    Drainage.. 59
    Estimate of coal 59
    General section across 50
    Roads... 48

HELENA BASIN—Continued.
    Topography... 59
    Varying dip of measures........................... 59
Helena Seam—Analysis of coal.................................72-73-77
    Dailey Creek basin................................. 108
    Dry Creek basin..................................... 76
    Eureka basin....................................... 69-70
    Helena basin....................................... 57-58
    Henryellen basin...............................27-35-36
    Lolley basin.. 86
    Overturned Measures................................ 98
    Sections of............................35-26-57-58-86-98
Henryellen Basin—Analysis..................................... 20-37
    Boundaries.. 21
    Drainage... 29-30
    Estimate of coal in................................. 36
    Roads... 22
    Section across...................................... 23
    Thickness of measures............................... 37
    Topography.. 29
Jones' Valley—Cambrian formation in....................173, 176, 178
    Eastern Red Mountain............................172-178
    Exceptions to general plan of structure............. 170
    General description................................ 170
    Knox Dolomite of..........................173, 175, 175
    South Highlands.................................... 173
    Western Red Mountain...........................172, 178
Knox Dolomite—Description of rocks of 150
Lancashire Mining Methods..................................... 120
Lemley Seam—Analysis of coal................................. 102
    Overturned Measures................................ 99
Lolley Basin—Area... 85
    Big falls... 89
    Boundaries.. 83
    Dip of measures..................................... 89
    Drainage.. 83
    Estimate of coal.................................... 85
    Roads... 84
    Topography.. 83
Little Mayberry Creek—Section along........................... 97
Little Pittsburgh Seam—Analysis of coal....................... 72
    Dailey Creek basin................................ 107
    Eureka basin..................................... 69, 70
    Helena basin....................................... 54
    Henryellen basin................................... 54
    Lolley basin....................................... 85
    Sections of.................................. 34, 35, 54
Luke Seam—Dailey Creek basin................................. 108
    Lolley basin.. 88
    Montevallo basin.................................. 92, 94

Mammoth Seam—Analysis of coal.................................. 31, 32
                Analysis of ash.................................. 38
                Analysis of coke................................. 38
                Henryellen basin................................ 26
                Section of...................................... 26
                Split in 26
Map of Cahaba Field—Account of its development............... 1, 2
Martin Seam, in Acton basin..................................... 45
Mining Methods.. 118
Monongahela Mining Methods................................118, 119
Montevallo Basin—Area... 91
                Boundaries..................................... 90
                Drainage....................................... 91
                Estimate of coal................................ 92
                Roads.. 91
                Topography.................................... 91
Montevallo Coal and Transportation Company..................... 94
Montevallo Conglomerate—Lolley basin........................... 88
                        Montevallo basin....................... 91
                        Overturned Measures................... 97
Montevallo—Change of direction of boundary of Cahaba Field near.. 168
Montevallo Seam—Analyses of coal.............................. 94
                Dailey creek basin.............................. 108
                Dry creek basin................................ 76
                Lolley basin.................................... 87
                Montevallo basin................................ 92
                Sections of.................................. 87-93
Montevallo Shales—Described................................... 148
Mountain Limestone... 156
                Sandstone bed in.........................156-157
                Quarried at Bangor, Blount Springs and Truss-
                    ville.. 157
Moyle Seam—Analyses of coal................................... 72
                Eureka basin................................... 72
                Helena basin................................... 54
Nunnally Seam—Acton basin..................................... 45
                Cahaba basin................................... 64
                Gould basin.................................... 81
                Henryellen basin................................ 24
                Section of..................................... 81
Orange Sand or Drift.. 159
Overturned Measures—Area..................................... 96
                Boundaries.................................... 95
                Dip of strata.................................. 97
                Drainage...................................... 95
                Estimate of coal............................... 96
                First mining in................................ 100
                Roads... 96
                Topography................................... 95
Oxmoor sandstone and shales................................... 155

INDEX.

Pelham limestone	152
Piney woods fault	88
Poole Seam—Henryellen basin	24
Section of	29
Post-Tertiary formations	158
Pratt's Ferry--Synclinal fold	167
Pump Seam—Helena basin	51
Henryellen basin	36
Quarry Seam—Dailey creek basin	107
Helena basin	55
Red Mountain or Clinton formation	153
Roup's Valley—Anticlinal near Vance's	180
Cambrian rocks in	178–79
Clinton strata	177, 178, 179, 180
Edwards' ore bank	179
General description	170
Knox Dolomite	177, 178, 180
McAshan mountain	177–178
Synclinal in, near Vance's	180
Salem Breccia	152
Section—General, across region shown on maps	161
Sections of Coal Seams—	
Acton seam, Acton basin	43
Beebe seam, Overturned measures	99
Black Fire Clay seam, Lolley basin	88
Black Shale seam, Helena basin	53
Buck seam, Helena basin	52
Cannel seam, Overturned measures	99
Clark seam, Dailey creek basin	107
Cooper seam, Overturned measures	99
Dodd seam, Overturned measures	99
Eureka Co's Slope seam, Acton basin	43
Gholson seam, Dailey creek basin	107
Gould seam, Gould basin	81
Helena seam, Dry Creek basin	77
Helena basin	57–58
Henryellen basin	35–36
Lolley basin	86
Overturned measures	98
Little Pittsburgh seam, Helena basin	54
Henryellen basin	34–35
Mammoth seam, Henryellen basin	26
Montevallo seam, Lolley basin	87
Montevallo basin	93
Poole seam, Henryellen basin	29
Pump seam, Henryellen basin	36
Shaft seam, Overturned measures	99
Thompson seam, Blocton basin	114
Three Feet seam, Overturned measures	99

SECTIONS OF COAL SEAMS—Continued.

Wadsworth seam, Cahaba basin	66
Eureka basin	71
Helena basin	50-51
Whetrock seam, Cahaba basin	66
Woodstock seam, Dailey creek basin	107
Shades' Valley—Underlying rocks	172
Shaft Seam—Overturned Measures	96
Sections of	98, 99, 100
Shute Seam—Cahaba basin	65
Helena basin	51
Silurian Formations	150
Smithshop Seam—Dailey Creek basin	107
Helena Basin	55
Strine seam, Dailey creek basin	108
Lolley basin	88
Montevallo seam	92-94
Sub-Carboniferous Formation—General description	155
Sub-Carboniferous Limestone—Varying thickness of	4
Thompson, or Underwood, or Conglomerate Seam—Analysis	72-115-116
Blocton basin	114
Dailey Creek basin	108-109
Eureka basin	69, 70
Lolley basin	85
Section	114
Three Feet Seam—Overturned Measures	99
Trenton Limestone	152
Tuscaloosa Formation	158
Underground Haulage	128-129
Underwood Seam—Same as Conglomerate and Gholson, analysis	72,115,116
Blocton basin	114
Dailey Creek basin	108
Eureka basin	69, 70
Lolley basin	85
Section	114
Valley Regions--Cambrian Formations	148
Choccolocco Shales	148
Classification of the rocks	146
Coosa Shales	148
Dearth of fossils	146
Distribution of the rocks in	159
Folds and faults in	141, 142
Folds not symmetrical	143
Formation of anticlinal valleys	144
Formations enumerated	138
Origin of the rocks	137
Paleozoic formations defined	146
Reversal of strata	142
Reversed thrust faults	142
Submergence of valley regions at different times	145

VALLEY REGIONS—Continued.

Thrust faults...	142
Variations in the rocks, with varying locality...	147
Wadsworth Seam—Acton basin...	45
Analysis...	60, 65, 73
Blocton basin...	114
Cahaba basin...	64
Eureka basin...	69–70
Helena basin...	50–51
Henryellen basin...	25
Sections of...	50, 51, 66, 71
Weisner Quartzite...	149
Whetrock Seam in Acton basin...	45
Cahaba basin...	66
Helena basin...	50
Section...	66
Woodstock Seam—(Same as Gholson), analysis...	110, 115, 116
Blocton basin...	114
Dailey Creek basin...	106, 109
Lolley basin...	85
Section...	107
Yeshic Seam—Dailey Creek basin...	108
Lolley basin...	86

www.ingramcontent.com/pod-product-compliance
Lightning Source LLC
Chambersburg PA
CBHW031815220426
43662CB00007B/658